ONCE UPON A DIGITAL STORY

A MODERN APPROACH TO AN ANCIENT ART

Written by Susan Tennant
with a contribution by Travis Faas

Indiana University — Purdue University Indianapolis (IUPUI)

cognella®
academic publishing

Bassim Hamadeh, CEO and Publisher
Michael Simpson, Vice President of Acquisitions
Jamie Giganti, Managing Editor
Jess Busch, Senior Graphic Designer
Angela Kozlowski, Acquisitions Editor
Luiz Ferreira, Licensing Specialist
Miguel Macias, Interior Designer

First published in the United States of America in 2015 by Cognella, Inc.

Printed in the United States of America

ISBN: 978-1-62661-410-9 (pbk) / 978-1-62661-411-6 (br)

cognella academic publishing

www.cognella.com 800-200-3908

This book is dedicated to Phil, Zach, Ethan, Amy, Julia, and Amelia.

Special thanks to Zeb.

CONTENTS

INTRODUCTION

The goal of this E-book is to understand the theory, concept, principles and development process of storytelling using digital media. Information included range from the history, conceptualization, development and technology of digital storytelling. The E-book explores the role of storytelling through time from traditional to digital. The book covers adapted, re-purposed and original stories. Specific subjects include interactivity, games, 2 & 3D animation, formats for different technologies and how to write digital fiction, non-fiction, transmedia and interactive story.

The book is divided into four sections:

Part 1 Covers the evolution and origins of storytelling

- History of storytelling
- Definition and process about how to create a digital story
- How and where digital stories are used
- The power of digital storytelling
- Mobisodes and webisodes

Part 2 Focuses on the developmental process

- Research and development
- The Hero's Journey
- Freytag's Pyramid
- Digital story structure
- How to write a digital story
- Character development

Part 3 Describes interactive stories

- Game development
- Interactivity
- Transmedia
- Script writing
- Storyboards

Part 4 Introduces the industry of digital storytelling

- Writing a game document
- Demo reels
- Making connections
- Dissemination
- Information about the field of digital storytelling and game development

Book Outcomes:

- Students will gain a deeper understanding of the history, culture, traditions and craft of digital storytelling.
- Students will learn through concept and practice how to construct a digital story.
- Students will learn how to develop linear and nonlinear stories.
- Students will learn about adaptation, re-purposing and original stories.
- Students will learn about interactivity, games and transmedia.
- Students will understand the importance of writing digital stories and what is expected of them if they were to enter the industry.

This book will give students the fundamental skills of how write a digital story and how to utilize multimedia such as audio, stereo graphics, video, 2 & 3D animation, motion graphics and interactivity. The book covers technical considerations such as format, dissemination including social media. There are numerous links embedded throughout the book that will enrich your understanding of writing a digital story.

1 Part One OVERVIEW

1. THE EVOLUTION OF STORYTELLING

A. HISTORY OF STORYTELLING

Humans have existed for more than 100,000 years. The earliest signs of literacy took another 5,000 years before mankind grasped the concept of drawing, then reading and writing. The act of storytelling has been a part of human history starting with the earliest caves (Figure 1.1) discovered in Altamira, Spain, and the Chauvet caves of Southern France. What do you see in the cave painting? What do you think the story is behind the imagery? From its origin, storytelling has been a means of communicating information. Initially early humans used storytelling as a method to transfer information about hunting, survival, environment, and food availability. The activity of storytelling throughout the ages has become a tool to communicate and capture the imaginations of people from young children to senior citizens. No one can say for certain who recited the first story or when. The Greek storyteller Homer, author of *The Iliad* and *The Odyssey*, is revered as the greatest of ancient Greek epics. The plot is about betrayal, love, revenge and war. The original Greek mythological story describes the Trojan War, a battle between the Achaeans (Greeks) and the Trojans after Paris of Troy fell in love with Helen and stole her from her husband Menelaus, King of Sparta. The war lasted nearly 10

Fig 1.1

Cave painting, Cantabria, Spain

Source: http://en.wikipedia.org/wiki/File:AltamiraBison.jpg.

years so the first part, *The Iliad* describes the Trojan War and how Helen was taken back after her lover Paris was killed and the prophesized death of Achilles comes true. *The Odyssey*, the sequel to *The Iliad,* is about Odysseus, King of Ithaca and his journey home after the fall of Troy. Two digital feature film based loosely on *The Iliad* and *The Odyssey* are *Ulee's Gold* and *O Brother, Where Art Thou?*

Fast forward in time to 1564 when William Shakespeare was born and raised in Stratford-Upon-Avon England. Shakespeare wrote theatrical stories that continue to entertain audiences worldwide. He wrote most of his best work between 1585 and1592. His narrative repertoire included Comedies: *All's Well That Ends Well, As You Like It, Cymbeline, The Comedy of Errors, Love's Labor's Lost, Measure for Measure, The Merchant of Venice, The Merry Wives of Windsor, A Midsummer Night's Dream, Much Ado About Nothing, Pericles, The Taming of the Shrew, The Tempest, Troilus and Cressida, The Two Gentlemen of Verona, Twelfth Night, The Winter's Tale*; Tragedies: *Antony and Cleopatra, Coriolanus, Hamlet, Julius Caesar, King Lear, Macbeth, Othello, Romeo and Juliet, Timon of Athens, Titus Andronicus*; and Historical plays: *1, 2, and 3 Henry VI, 1 and 2 Henry IV, King John, Henry V, Henry VIII, Richard II, Richard III*. In total Shakespeare wrote 37 plays. His narratives continue to be the basis for stories such as 1956 *Forbidden Planet* based on *The Tempest* and *Ran* (1985) based on *King Lear*. An excellent way to learn how to write a story is to read the works of Shakespeare and classics by John Steinbeck, George Orville, J. R Tolkien, Ernest Hemmingway, F. Scott Fitzgerald, Ray Bradbury, Mark Twain, John Updike and Kurt Vonnegut, etc. I recommend that you read and familiarize yourself with the writing style and character development of these authors and others. There are so many authors to recommend. Some of my favorite include: Margaret Atwood, Louise Erdrich, Ian McEwan, Jim Harrison, John Irving, Toni Morrison, Alice Walker and Mark Helprin's, who wrote A Winter's Tale, which was adapted into a feature film in 2014. Contemplate how the story has evolved from the earliest means of expression; an example is the 3D documentary *Cave of Forgotten Dreams*, based on the discovery of the Chauvet cave painting in 1996 by Eliette Brunel-Deschamps, Christian Hillaire, and Jean-Marie Chauvet for whom it was named.

A function of storytelling has been to connect the past, present, and future into narratives that educate, entertain, and inform an audience. The genres of stories are wide and varied, ranging from comedic animations to epic adventures, entertainment to education; based on fact or fiction; the act of storytelling has played an important role in the tapestry of human history. Stories help all of us to understand the world and its mysteries and complexities.[1] Watch this video about the "History of pretty much everything," to bring into perspective the history of civilization as a short animation and an introduction to this book.

The dynamics of storytelling changed with the invention of the binary code system that converted analog text, visuals and audio into two binary units, 1 and 0. This innovation, discovered in 1679 by Mathmatician, Gottfried Wilhelm Leibniz, has enabled modern authors to create, edit and distribute their story more efficiently as opposed to typing or hand writing them. Trademarks of digital stories are: can be created and edited entirely on a computer, it can be viewed online anywhere and anytime, downloaded, shared, and archived. Stories that are available online in the form of television, mobisodes, webisodes, feature films, games, and social media can also be interactive. Keep in mind that not all digital stories are interactive. Interactivity is a method used in some digital stories.

B. ADAPTATION, REPURPOSING, AND ORIGINAL DIGITAL STORIES

The evolution of traditional story to digital story is constructed through these three perspectives:

I. Adaptation

The purpose of adaptation is to present an updated version of the story so the audience has a different experience from the original. An example is the evolution of Batman from television to the silver screen. The story has been adapted and enhanced because of digital effects. A number of children's stories are adapted remove hyperlink, retold from one medium to another, such as the European fairy tale *Little Red Riding Hood* by the Brothers Grimm. This story about a young girl and a big bad wolf has

pg 9

Fig 1.2

Little Red Riding Hood by Carl Larsson
Source: http://en.wikipedia.org/wiki/File:Carl_Larsson_-_Little_Red_Riding_Hood_1881.jpg. Copyright in the Public Domain

been retold and adapted into a Disney animation in 1922 and in 2009 a feature film, *Red Riding Hood* (Figure 1.2).

Typically the adapted story will retain the context of the original. The transformation will change depending upon what media is used. A story adapted from print to film will have the same plot and characters as the original but might be shortened. In the same regard an adapted story can be enhanced by using digital effects and multimedia. Feature films and television stories are heightened because of multimedia-CGI (Computer generated imagery), Dolby sound, animations, 3D and special digital effects. Stephen King's "The Body" is an example of a short story that was adapted into the film *Stand By Me* directed by Rob Reiner. In game play, the adaptation is different because the game has many subplots from which the player can choose to follow. In video games based on books such as Harry Potter and Alice in Wonderland the transformation needs to take into account that the story is nonlinear, meaning the plot can go in a variety of directions depending on choices made by the end-user. There are many books adapted from print to film, performance, television, games and video. *Superman*, the comic, ran the gamut from comic book to clothing, television, feature film, toys, and video. Multimedia developed by One 3 Production http://one3productions.com/o3p/One_3_Productions_-_Every_Moral_Has_a_Story_-_Transmedia_Creators_and_Consultants_Los_Angeles.html is a business that produces digital stories for a variety of delivery platforms ranging from film to augmented reality games. The question to the writer of a digital story is what keeps the audience's attention? An example of an adapted story, is the 2010 movie *Scott Pilgrim vs. the World,* World. This digital story originated as a comic. The story is based on the character, Scott Pilgrim, who must fight seven "evil ex-boyfriends" of the young woman he has fallen in love with. Another is the comic adventure *300* written and illustrated by Frank Miller. The comic based on the Persian Wars was adapted to The *300 Spartans* in 2006 and in 2014 retitled *300: Rise of an Empire*.

II. Repurposing

Repurposing a story is defined as taking the narrative and preserving the content but upgrading the method of distribution (platform) to the most current technology. Examples include VHS to DVD followed by streaming over the Internet. Repurposing is a method of preserving and archiving stories so they can be of benefit to anyone from researchers to individuals and live on from one generation to the next. Through the repurposing process, the story will not change; only its digital format (.Mpeg4, .Mp3, etc.) and platform (Blu-ray, 3D) change. Gordon Moore stated that every two years the processing speed, capacity, and memory of technology would increase and even double in capacity. Repurposing is Moore's law put into practice. Historically "repurposing" is a term used in the television industry when an original television series ends. Repurposing a TV program became a standard to extend the show's playtime as a series, rerun or spinoff.

Fig 1.3

Audience with 3-D glasses.
Source: http://en.wikipedia.org/wiki/File:The_National_Archives_UK_-_WORK_25-208.jpg. Cleared via Open Government Licence v1.0.

III. Original Digital Stories

Is there such a thing as an original story? Many writers struggle to glean that novel idea. There is no true answer for this question. Some stories have unique plots that twist and turn unexpectedly, such as Memento, Fight Club, Pulp Fiction, Pleasantville, Inception, The Prestige to name a few. The transformation of the story from the campfire model to the silver screen and the current technology of watching stories on mobile devices haven't changed the demand for narrative: if anything it has given the audience more choices for accessibility. A creative use of technology is the online blog Dr. Horrible's Sing-Along, starring Neil Patrick Harris. This digital 2008 musical comedy dramatic online miniseries is a three act narrative that depicts Dr. Horrible's escapades with his antagonist, Captain Hammer and his love interest, Penny. In 2008 the story won the People's Choice award for "favorite Online Sensation."

Digital stories such as Avatar and Gravity, are filmed in 3D to enhance audiences depth perception and give the viewer a unique experience. Narratives filmed in 3D create an immersive experience for the audience by creating an illusion of depth. The audience wears stereoscopic glasses (Figure 1.3) which creates the 3D effect. One of the earliest 3D films, which premiered on June 10th 1915 was the silent film *Power of Love* by Edwin S. Porter.

The weaving of a story either fiction or non-fiction involves building a plot that is intellectually and emotionally meaningful. As we have learned, storytelling in the digital age retains characteristics of traditional storytelling. These include the ability to entertain, inspire, inform, train, and/or educate. Storytelling hasn't changed in form, but has evolved through the advances of technology. Some things to in the development of a digital story is the value of technology in the creation and production of the digital story. What makes Shakespeare's stories timeless? The question is what value does the technology bring to the story? Would it be more effective to view it on a large-format screen or on a personal mobile device; in 2D or 3D or game or animation? Classical stories connect to audiences because the premise is enduring. Classics reach a deeper level of investment that lingers long after the story is told. Any great story must infuse the reader with a sense of emotional connection that captures a universal theme about human perspicacity. The bottom line is to make your audience care. Shakespeare wrote stories with universal themes. Romeo and Juliet, a story about love, betrayal, and family conflict reportedly written in 1594 or 1595, has been recreated many times; here is a partial list of some film adaptations:

- *West Side Story* (1961)
- *Romanoff and Juliet* (1961)
- *Romeo and Juliet* (1968)
- *Cocktail* (1988)
- *Titanic* (1997)
- *Romeo and Juliet* (2013)

It is human desire to be transported into the plot of a story. The journey can be transformational. it is important to keep in mind that technology cannot enhance a story if the premise is poorly constructed. No matter what media or technology is used, the bottom line is the story has to be good. We will get into that subject in Part 2 of this eBook as you learn how to develop a digital story. An amusing history of storytelling is depicted in this video on YouTube entitled, Whiteboard History of Storytelling.

C. WHAT IS DIGITAL STORYTELLING?

The word **story:** ***noun*** **\'stȯr-ē\—from Latin *historia*,** Archaic: History meaning: of, relating to, or being data in the form of especially binary digits <*digital* images> <a *digital* readout>; *especially*: of, relating to, or employing digital communications signals <a *digital* broadcast> — compare. Relating to an audio recording method in which sound waves are represented digitally (as on magnetic tape) so that in the recording background noise is reduced.

Throughout human history, storytelling has been used to share knowledge, wisdom, and values. The evolution of stories has taken many different forms from print, e-book to 3D feature films. Digital stories use many traditional story themes: romance, comedy, drama, horror, action, adventure and suspense. Watch this TED talk by Joe Sabia: The Technology of Storytelling. The use of technology in storytelling must not be selected because it is the most current but rather it is the most compelling method to tell and/or enhance the story. A novel written across 300 pages of paper and a novel conveyed across 300,000 separate tweets are not really any different.

Fig 1.4

Picture of the original Gutenberg press.
Source: Skip Comer.

D. PROJECT GUTENBERG

The format of the story changed dramatically with Johannes Gutenberg's invention of the printing press and moveable type in 1450 (Figure 1.4). He tested his moveable-type machine by printing a Latin book on speech making. When this endeavor was successful, he embarked on his most famous project, the printing of The Gutenberg Bible. Prior to this invention, books were hand printed on parchment and papyrus. The hand-written book took hours to create and so, few copies were produced. Early books had a short shelf life due the fragility of the medium it was written upon. The first books originated from Gutenberg's hand press, began the process of mass production. The early book-making process involved spreading ink over handset raised block letters and rolling a sheet of paper through a printing press, which produced a page of text and/or drawings. The pages were bundled into a hand-stitched book that could be shared with other readers. This revolutionized storytelling for centuries.

E. INVENTION OF THE EBOOK

The production of storytelling evolved largely due to the advancement of technology. Starting with the inventions of radio, silent movies, movies with sound, and video-to-online formats, the attraction to good narrative in the course of human history remains undiminished. In 1971, Michael S. Hart invented the first electronic book, the eBook. This came about by Hart experimenting with the ARPANET network during a time when the Internet was very young. Acquiring a free copy of the U.S. Declaration of Independence, he typed this important document into a computer and put it on the Internet (World Wide Web) for anyone to download. With this action, he became a pioneer in the history of digital storytelling. Hart's importance also includes the fact that he founded Project Gutenberg. This project was started in 1971 as a celebration of the invention of the Gutenberg printing press. The significance of the Project Gutenberg collection is that it offers selections of free stories whose copyrights have expired. The depository consists of over 42,000 eBooks. The story Youth by Isaac Asimov is a good one to start with from this treasure chest. In July 2011, shortly before his death, Hart wrote, "*One thing about eBooks that most people haven't thought much is that eBooks are the very first thing that we're all able to have as much as we want other than air. Think about that for a moment and you realize we are in right job.*"[5]

F. ADVANCEMENT OF TECHNOLOGY

The influence of technology has enabled the story to be adapted for a number of digital devices such as personal computers and mobile phones. Digital storytelling is ubiquitous, so anyone who has connectivity to the Internet can create, produce, and disseminate their story. Most everyone wants to tell a story. The reason is universal; everyone has a story to tell. In the same regard, most everyone spends time engaged in the activity of watching, listening to, reading, or experiencing a story in some form or another. The action of storytelling ranges from a solitary activity to an immersive and interactive one. Over time the story has changed with the advancement of technology. Prior to the 1970s audio and visual recording devices produced analog results with few options to reduce noise from a sound recording or a mistake from 8mm or 16mm film. Engineers began exploring digital technology and converting the media into binary code. The first digital audio recorder was the Sony PCM-1 in 1977 (Figure 1.5). The first digital audio recorder was developed by NHK (Japan Broadcast Company) in 1967 and fully used by the BBC, one of the first broadcast companies to use digital audio technology to transmit from their studios to the antennas. This significantly boosted quality. Sony then led the way with technology, first writing PCM audio to Betamax, then to DAT and the CD in 1974, though the CD wasn't really prevalent until the final demise of tape. The tech head of NHK, Heitaro Nakajima, who was integral in inventing the early digital audio recording techniques, was hired by Sony to develop the digital audio department (1971). According to the Library and Archives of Canada, early digital recordings were cumbersome due to the process of changing analog information into binary code. The audio sound wave consisted of sampling the sound wave frequency, measuring the amplitude, and assigning binary values to each wave. In digital format the sound distortions and noise factor could be reduced and eliminated in post-production. The VTR (video tape recorder) captured images into digital format that was recorded onto magnetic tape. The first video camera to have sound was the Sony BMC100P Betamax in 1984 and the first camcorder using the VHS format was the JVC GR-C1.

Digital recordings have gone through many formats. Steven Sasson, an engineer for Kodak, invented the first digital camera in 1975 (Figure 1.6). In a Kodak blog post written in 2007, Sasson explains how the digital camera operated, "*It had a lens that we took from a used parts bin from the Super 8 movie camera production line downstairs from our little lab on the second floor in Bldg 4. On the side of our portable contraption, we shoehorned in a portable digital cassette instrumentation recorder. Add to that 16 nickel cadmium batteries, a highly temperamental new type of*

Fig 1.5

Sony camcorder

Fig 1.6

First digital cameras by Kodak and Steven Sasson

Source: Kodak

CCD imaging area array, an a/d (analog to digital) converter implementation stolen from a digital voltmeter application, several dozen digital and analog circuits all wired together on approximately half a dozen circuit boards, and you have our interpretation of what a portable all electronic still camera might look like."[6]

The Internet has become a destination to exchange and download audio and video recordings. Compression changes have reduced the recording sizes so media can be downloaded and uploaded quickly. The popularity of Mp3 for audio and Mpeg4 for video opened up the market of personal recording and player devices. Manufacturing of these digital recorders became competitive and the marketplace offered a wider variety of choices. When Apple introduced the first iPod in 2001, their motto was to put "1,000 songs in your pocket." Apple launched the iTunes music store in 2003, which offered consumers 200,000 songs for 99 cents each along with a thinner iPod that had the capacity to hold 7,500 songs. That same year Apple's iTunes sold one million iPod's in the first week. Using a personal digital device (PDA), an individual could listen to music or view videos at their leisure. Any consumer could also create and produce his or her digital story. [7] These inventions created opportunities to tell stories and share them with others through a global network. The power of the story can be only as good as the story. Adding audio, special effects, visual elements, and/or interactivity should be used to enhance the story, not just to give it entertainment value. These elements add dimension to the story that could not have been achieved through traditional methods. The inclusion of interactivity offers the end user opportunities to participate directly with the story.

EXERCISES FOR STUDENTS

a) Read the short story *Youth* by Isaac Asimov from Project Gutenberg.
b) How does the illustration influence the narrative.
c) Discuss the history of storytelling.
d) Discuss adaptation, re-purposing and original type stories. Find examples.

e) Create a drawing of a cave painting and have a classmate interpret the image.
f) Discuss why are stories important to our culture and understanding of the world.
g) Discuss how technology has influenced, advanced, storytelling and find examples.

2. THE ART OF DIGITAL STORYTELLING

A. WHAT MAKES A GOOD STORY?

Story ideas come from anywhere and everywhere. Look at your own world and the people in it. Are any of them memorable characters? A principle of digital storytelling, is that the story must be engaging, entertaining, educational, or informative to keep the audience's attention and make them care.

Fig 2.1

Legs sticking out of water barrel

I. Where Do Ideas Come From?

Stories surround each of us every day. Does this mean everything humans do or say is a story? No, as a writer you need to learn to filter ideas. Begin by conducting research, reading books, watching films, and going to the theatre to learn from master storytellers. The most important thing to keep in mind when writing a digital story is to use the media appropriately. No amount of media or technology can improve a story that is poorly written. To see with the eye and the mind is a good start towards ideation. "*The greatest thing a human soul ever does in the world is to see something, and tell what it saw in a plain way,*" (quote by John Ruskin, art patron and leading art critic of the Victorian Period (1819–1900). A writer needs to be an observer and take in his or her surroundings. Use your imagination to build on an idea. Write it down or rather type it into your electronic device. The more you write, the better your work will

Fig 2.2

Woman reading a book with Portrait of Vincent Van Gogh

Fig 2.3

Person posing as Statue of Liberty

get. What is the story behind the image of legs sticking out of the water barrel (Figure 2.1)? How come the legs only have one boot? Could you enhance this scene, bring it alive by using digital effects/technology? Create a story about the origin of the mannequin; who put it there; why is it there? Then review the scene in (Figure 2.2) and create a scenario of why is the person sitting in front of a Van Gogh painting? Is she waiting for someone? What is she reading? The bottom line is, pay attention so you can develop concepts based on your observation(s).

Stories often originate from what the writer knows. There is no better foundation than to write about something you know or care about such as family, interests, and hobbies. A writer examines all that occurs in his world. Observe the statue of Liberty (Figure 2.3) who is dressed in white and standing on a bridge. As a digital storyteller my suggested method of creating a story is to contemplate these questions:

1. Why is the person dressed up as the Statue of Liberty?
2. What is the story about?
3. What is the POV of the story?
4. Where does the story take place?
5. Who are the main character(s) and what are their personalities like?
6. When? What time of day, week or year does the story take place?

These observations are clues that add context to the story. The scene (Figure 2.4) captures 4 people in distinctive colorful outfits. What are they doing? Why are they dressed so vibrantly? What is their connection? These are questions that can be utilized as you build your concepts for digital stories. The mystery author Dan Brown, in his book *The Da Vinci Code*, keeps the audience interested by feeding them clues. Become acutely attuned to your surroundings and take notes about the most interesting, obvious, and even trivial things you observe. Watch this TED Talk by Andrew Stanton. Be forewarned that there is a curse word at the beginning of this video. Stories originate in many ways and a good place to start is to review any of the timeless classics of book, radio, television, film, and games. Think about "what if," "why that"—the best place to start is to write about something that interests you.

Come up with something novel such as writing about superheroes who have useless super powers such as being able to fly only up; having laser vision that can only read people's minds while trying on pants; having super powers that only work in the rain. Recognize your perception of the world and learn to pay attention to people and activities around you. Work on concepts and in time you will come up with that nugget of an idea. Keep a notebook by your bed and get in the habit of writing down your ideas, before you go to sleep or when you awaken. Make a list of your ideas, then edit till you have a few topics that are interesting and narrow that down to one. Ideas come from observations, researching, and watching and reading many different types of stories. Feedback is important and so is constructive criticism, so discuss your ideas with anyone you're comfortable sharing them with. Be a people watcher and observe people doing whatever they do and think about who they are, what they did last night, their background, etc.

Fig 2.4

People at a café in Amsterdam

Studying how people go about their daily activities is an excellent method for creating a character(s) biography for your story. One of my favorite concept development activities is to people watch. Take a walk where people gather such as at a park, popular street, a shopping mall or sporting event. Observe people and their activities. Look at the picture (Figure 2.5) and create a scenario by asking questions such as why is the man sitting on a couch in the middle of the street under a parking sign?

Fig 2.5

A man on a couch

Take a digital mobile device such as your smart phone or digital camera and capture images that you find interesting. Go out to dinner and watch people near you and conjure up scenarios such as is this a first date or are the couple married? What are their occupations and what are they discussing? Take the idea one step further and embellish the story using multimedia, digital effects. Animate the restaurants waiters and produce the story as a PSA (Public Service Announcement) for employing cartoon characters. See how a story can grow from a simple observation? Be a people watcher and observe people doing whatever they do and think about. Who are they? What are they doing? Where do they come from? Why are they so interesting that they capture your attention? Can the scene be a basis for a story? Build upon the people you observe

Fig 2.6

Aristotle
Source: http://en.wikipedia.org/wiki/File:Francesco_Hayez_001.jpg. Copyright in the Public Domain.

and develop their biographies. If your characters are interesting then the story will build itself.

B. ARISTOTLE AND FREYTAG'S STORY STRUCTURE

Stories, according to Aristotle (Figure 2.6), are based on cause and effect that link the incidents together. He wrote that good stories include an element of change in which the character(s) evolve. Aristotle also wrote that a good story must include the following elements:

1. Plot—sequence of events of which the story is composed
2. Character—individuals within the story
3. Theme—the controlling idea/central insight
4. Point of View—whose eyes the story is told through
5. Symbol—means more than one thing
6. Irony—incongruity, used to suggest the difference between appearance and reality, between expectation and fulfillment
7. Conflict/tension—adds emotional value
8. Goal—what the character(s) achieve at the end

In dramatic stories such as a tragedy, the idea is to arouse emotions from the audience such as rooting for the protagonist (hero) and loathing the antagonist (villain). Aristotle wrote that a good story culminates with release of emotions. An example is a catharsis, defined as a "purifying or figurative cleansing of the emotions, especially pity and fear," described by Aristotle as an "effect of tragic drama on its audience." What is known about Aristotle is that he was born sometime around 384–322 B.C. (62). He was a Greek philosopher (student of Plato's at The Academy in Athens from 18 to 38 years of age). He taught King Philip II of Macedonia's son Alexander, who became known as "Alexander the Great," and conquered all of Greece and the Persian Empire.

In 334 B.C. Aristotle founded his own school in Athens, the Lyceum. He fled in 323 B.C., fearing for his life, when the Athenians came to charge him with impiety (acting without reverence to the gods) just after Alexander died. Aristotle's connection to storytelling was his contribution of story structure. The Freytag Pyramid uses Aristotle's model to illustrate this three-act story structure:

1. Act I: Exposition (beginning, thesis)
I. Introduces the characters (main and supporting plots)
II. Introduces the location/setting/tense
III. Introduces the POV – Point Of View
IV. Establishes the themes, goals, tasks, outcome
V. Conflict, obstacles, problem, adversary, suspense, tension
VI. Ends with reversal, or new task/adventure, setback or change

2. Act II: Climax (middle, antithesis)
I. Raises the stakes
II. Introduces subplots, subordinate characters
III. Creates a reason to continue to find resolution
IV. Catharsis, purging of emotions
V. Sets up the antagonist and protagonist predicament
VI. Ends with more information, higher level of crisis

3. Act III: Denouement (end, synthesis)
I. Problem intensifies
II. Subplots resolve into main plot
III. Brings resolution to the antagonist and protagonist predicament
IV. Meets goal of story

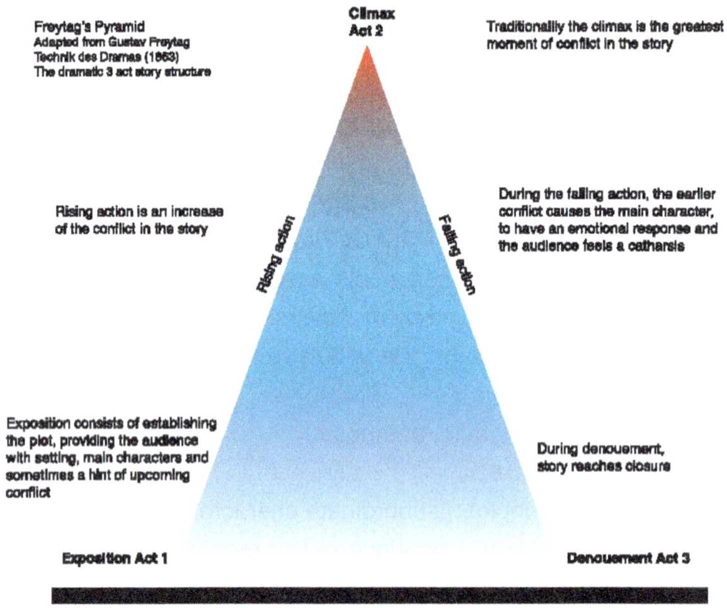

Fig 2.7

Gustav Freytag Story Structure Pyramid

In 1863 Gustav Freytag created a pyramid (figure 2.7) that was a visual representation of Aristotle's three-act play.

C. *STAR WARS* AND BEYOND

A generation grew up on the epic adventure Star Wars. The first movie was shown to the public in 1977, followed by the final film of the trilogy in 2012. Each film of the trilogy was released in three-year intervals. There are multiple trilogies, two released and a third in development. The first film of the first trilogy (Episodes 4,5,6) was shown in 1977, 1980 and 1983. The last film of the second trilogy (Episodes 1, 2, 3) were shown in 1999, 2002 and 2005.

According to wikipedia.org, "All six of the main films in the series were also nominated for or won academy awards. George Lucas in Love is a comedic interpretation of how George Lucas as a student got the concept for *Star Wars*.

Developing a series of digital stories is a much different ball game than the development of one story, whether it is digital or not. Some of the blockbusters that are part of this genre include: *Star Trek*, *The Tolkien* trilogy, *The Harry Potter* series, *The Hunger Games*, and *Star Wars*. Each of these stories must be able to stand as a complete story on its own merit and also connect to the next part of the narrative. In each story the content must be strong enough to maintain the audience's attention. Therefore it is important for the main character(s) to change over time. If they do, according to Aristotle, the story will be successful.

EXERCISES FOR STUDENTS

a) Compare and contrast digital and traditional stories.
b) Take pictures with digital camera and create stories about the pictures.
c) Discuss the TED talks.
d) Discuss what makes a good story.
e) Interview someone from your neighborhood and create a story about story about that person.
f) Create a digital story about something interesting in your neighborhood such as historic landmark.
g) Conduct a writing workshop and create a three act play using Aristotle's structure.

3. THE POWER OF STORIES IN THE DIGITAL AGE

A. STORIES HAVE POWER: THE SHOAL FOUNDATION

Fig 3.1

People of Paros, Greece

Technology can be used to digitize stories and build an archive for future generations to learn from the past. Collecting stories about farming and the effect of tourism on a community such as the one depicted in "People of Paros, Greece (figure 3.1) became the focus of the video Food of the Past, Nourishment for the Future, produced to bring attention to this beautiful Cycladic Island. These human-interest stories have the potential to enrich a global audience's understanding and appreciation of this beautiful island's story, culture and traditions. Another example is the USC Shoal Foundation created by Steven Spielberg to tell the story of U.S. citizens who survived the Holocaust and other genocides. This digital archive contains 52,000 audio/visual testimonies of survivors and witnesses. An example of the power of storytelling, *Scientific Adam and Eve*, traces the connectivity of human beings through the use of animation, special effects, and motion graphics.

Many museums use current technology such as a smart phone for self-guided tours of their exhibitions (Figure 3.2).

Fig 3.2

Prototype of Museum Digital Tour

This technology allows a patron to listen to stories about the artifact they are viewing. Museums such as the Smithsonian Museum in Washington, D.C also offer a website for end users to learn more about artifacts in their collections.

At the Children's Museum, in Indianapolis, Indiana, there is an interactive story on permanent exhibition, *The Power of Children: Making a Difference*. The story is about the lives of Anne Frank, Ruby Bridges, and Ryan White, three children who overcame enormous obstacles in their young lives. Each story includes artifacts that were part of the children's lives. Anne Frank's story and display contains replicas of the secret room she and her family hid in during the Nazi reign; the classroom where first-grader Ruby Bridges spent a school year alone with her teacher Mrs. Henry because she was the first African American child to attend an all-white elementary school; and Ryan White's bedroom filled with memorabilia from friends and celebrities who took on Ryan's cause to stay in public school after being infected with HIV/AIDS.

B. CHARACTERISTICS OF TRADITIONAL, DIGITAL, AND INTERACTIVE STORIES

Characteristics of **traditional** storytelling:

1. Are pre-constructed (elements can't be changed)
2. Have a linear plot (usually told from beginning to end)
3. Author/writer is the sole creator
4. Are experienced passively
5. Have one unchangeable ending

Characteristics of **digital** storytelling OFTEN:

1. Blur fiction and reality
2. High definition, mobile accessible
3. Include digital media
   - 3D, animation, computer-generated graphics (CG)

4. Can change POV
5. Can include elements of play and interactivity

Characteristics of **interactive** stories:

Digital stories that are interactive have a different structure (Figure 3.3) than the traditional story. In an interactive story, the end user controls the path of the character and how the narrative plays out.

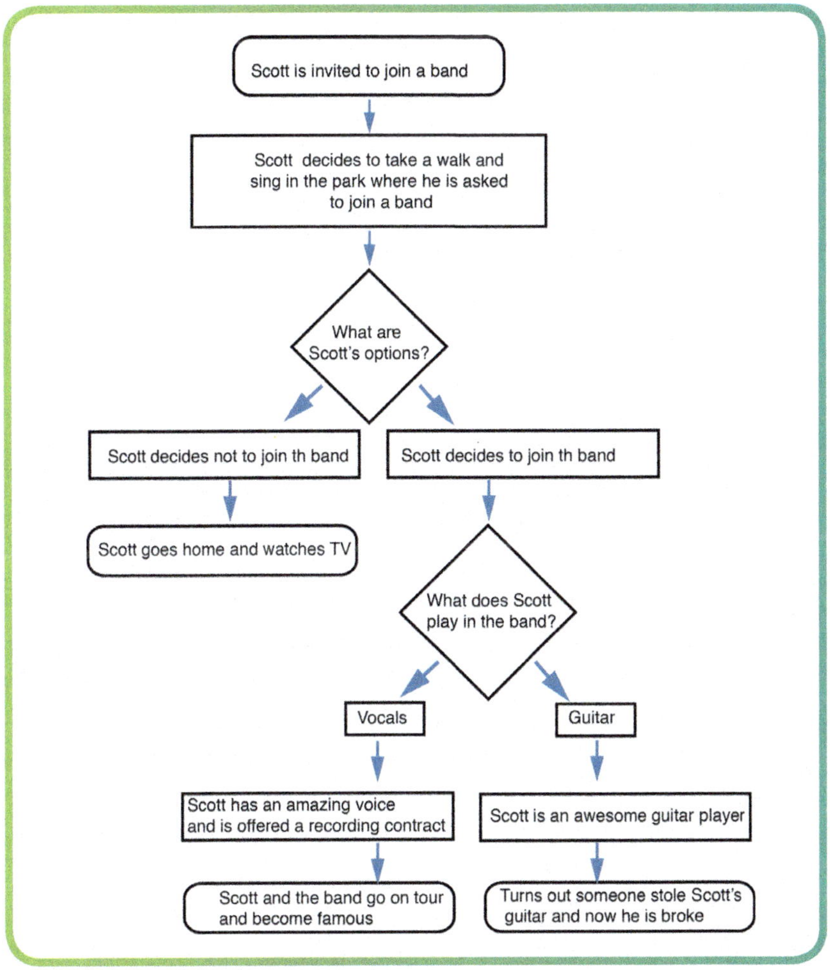

Fig 3.3

Flowchart for a Choose Your Own Adventure interactive story

Using technology interactively allows two or more people to engage in storytelling over a global communication system. Interactive stories provide end users different paths to select. Here are some characteristics of interactive stories:

1. Use digital effects
2. Have a system of reward/penalty
3. Are multisensory
4. Incorporate AI (artificial intelligence)
5. Offer a shared experience
6. Manipulate time and space
7. Put users through challenges and tests
8. Contain characters—can be controlled by the user or a computer and may be synthetic with AI
9. Equip user with controls and input devices
10. Deeply immersive—pull user into story
11. Participatory—participate in the story
12. Navigable—user makes own path

C. IMMERSION

A digital immersive story is defined as being surrounded in an engrossing environment to view or participate in the action of the narrative. Stimulating multiple senses can draw the end user into the story and create an environment in which he or she connects to the story, resulting in a more impactful narrative. Immersive stories such as *Avatar,* shown in 3D, take the viewer into the narrative's virtual space. Objects and people appear to be coming out of the screen, close enough to touch. Another type of immersive story takes place in the C.A.V.E. (Computer assisted virtual environment). This is an immersive environment that uses multiple projectors to create a three-dimensional space. The C.A.V.E. projects visual effects onto walls, ceiling, and floor space to put the viewer directly into the virtual space by using stereographic goggles and a joystick. Some characteristics of immersive stories include multisensory elements that use sight, sound, and touch (haptic technology) to enhance the user experience. An immersive story envelops the user, takes them away from the real world and into a 3D virtual world.

EXERCISES FOR STUDENTS

a) Write a story that incorporates multimedia.
b) Go to a museum and discuss your experience and how museums tell stories.
c) Research community based stories.
d) Discuss the characteristics of traditional, digital and interactive stories.
e) Create a digital tour of your neighbor, local attraction, etc.
f) Discuss how immersion changes/influences the story.

4. DIFFERENT TYPES OF DIGITAL STORIES

A. NARRATIVE AND NON-NARRATIVE

Narrative. Storytelling can be classified into two categories: fiction and non-fiction. Fictional stories tell a tale of imaginary events. These can be in the form of comedy, drama, romance, and/or tragedy. One fictional story—*Alice's Adventures in Wonderland,* later changed to Alice in Wonderland—was written originally as a children's book In 1865 by the English author, Charles Lutwidge Dodgson, under the name Lewis Carroll (Figure 4.1). This classic narrative has been adapted successfully from print to animation to feature films.

The book and movies unravel the story about Alice falling down a rabbit hole. She discovers a different world than where she came from. The adventure is filled with creatures of delight, confusion, and terror. First produced as an animated fantasy by Walt Disney in 1951, the story has been adapted three more times, most recently in 2010 starring Johnny Depp, in which Alice as an adult returns to Wonderland.

Non-fiction stories are based on reality and actual events. Biographies and documentaries fall under the heading of non-fiction stories. Digitizing these types of stories is a way of preserving them for future generations. Non-fiction digital stories can be useful in telling about an event that occurred in

88

"The Queen of Hearts she made some tarts
All on a summer day:
The Knave of Hearts he stole those tarts,
And took them quite away!"

"Now for the evidence," said the King, "and then the sentence."

"No!" said the Queen, "first the sentence, and then the evidence!"

"Nonsense!" cried Alice, so loudly that everybody jumped, "the idea of having the sentence first!"

"Hold your tongue!" said the Queen.

"I won't!" said Alice, "you're nothing but a pack of cards! Who cares for you?"

At this the whole pack rose up into the air, and came flying down upon her: she gave a little scream of fright, and tried to beat them off, and found herself lying on the bank, with her head in the lap of her sister, who was gently brushing away some leaves that had fluttered down from the trees on to her face.

Fig 4.1

Alice's Adventures in Wonderland by Lewis Carroll

Source: http://en.wikipedia.org/wiki/File:Alice_par_John_Tenniel_30.png.

the past and sharing with a universal community. Ken Burns, a documentary filmmaker, has made films about baseball, the Dust Bowl, America's National Parks, jazz, and a seven-part mini-series shown on television in 2007, entitled *War.* His films are impeccably researched and rely on factual information. In the documentary series *War*, Burns combines interviews of veterans with actual footage shot during WWII. Produced by Vice President Al Gore in 2006, the documentary *An Inconvenient Truth* uses digital animation and special effects in conjunction with real footage to tell the story about global warming.

Non-fiction digital stories can be used to tell the story about people, places, or events. An example is the story about the nonprofit organization, *The American Red Cross*. In 2012, hundreds of cameras were sent to the people helped by the Red Cross. The footage was returned to the Red Cross, who then edited the clips into one digital story. *Food, Inc.* is a non-fictional account of the inside of America's corporate food industry that is told through the stories of farmers affected by the rise in production of genetically modified food. This documentary in particular is compelling due to the truism of the story.

B. GAMES AND INTERACTIVITY

In game design the story is essential to the activity of playing the game, whether the game is a first-person shooter, a role-playing game, or multiplayer. With the advancement of technology, the premise of the story is an essential part of the gaming experience. The stories developed into games cover a variety of subjects from health and sports to action adventures. Some of the earliest arcade games, such as Pong, didn't have a story structure, but was rather a game of tennis that matched player to the computer. One of the first story-based video games, Donkey Kong, developed by Nintendo in 1981, involved the main character maneuvering through obstacles as he collects rewards. In the story, Mario, originally known as "Jump man," tries to rescue the lady in distress (Pauline) from a large ape named "Donkey Kong." As story-based games became the benchmark of success, the experience involved richer and more intricate scenarios.

Merriam-Webster online dictionary defines "interactivity" as:

1: mutually or reciprocally active

2: involving the actions or input of a user; *especially*: of, relating to, or being a two-way electronic communication system (as a telephone, cable television, or a computer) that involves a user's orders (as for information or merchandise) or responses (as to a poll).

C. INTERNET STORIES

I. Webisodes

The webisode is a combination of the words "web," "Internet," and "episode." One of the earliest webisodes, the *7th Portal,* was produced in 2000 by the legendary comic book writer and producer Stan Lee. His intention was to extend the life of the *7th Portal* comic series with an online version of the superhero story. Webisodes are stories accessible only on the Internet. Developed in the 1990's, webisodes became very popular because they could be downloaded or streamed onto a personal computer. The webisode narrative covers a variety of genres from action adventure, comedy, drama, fantasy, romance, and science fiction to thrillers. Webisodes are adaptations of stories first delivered on broadcast television. The webisode is a series of stories formatted for the Internet that ties together multiple episodes from a central plot. The story structure of a webisode is the same as any narrative. The webisode script is structured using Freytag's three-act play model. Each chapter in the series must be a story unto itself, with enough content to keep the audience's attention for the next webisode. The main difference between a television series and a webisode is the time factor, screen size, and resolution. Professional broadcast stories are traditionally broken down into sequences or chapters and delivered to the audience over a dedicated timespan. The television time is restricted to a 30-minute timeslot; actually 22 minutes with commercials for a half hour show, and one-hour stories, usually 45 minutes with commercials. Webisode stories are usually shorter: the average length is 5 to 10 minutes.

Broadcasting networks such as NBC have taken advantage of webisodes to extend the life of their television series after the original stories

Fig 4.2

Mobisode

were no longer available on primetime. The advantage of watching a webisode is they are shortened stories to be viewed at any hour or location as long as the end user is connected to the Internet.

II. Mobisodes

A mobisode is a combination of the words "mobile", "Internet" and "episode". The U.S. broadcasting company Fox trademarked the term in 2004. A mobisode is similar to webisodes in that both are typically shortened Internet versions of the original television story. Mobisodes were developed for a small screen (Figure 4.2), which measures 4" H × 2" W. A mobile unit's screen resolution is smaller than a personal computer's screen or HDTV, so the digital story must utilize the technology efficiently and effectively. Mobisodes timespans are generally 30 seconds to 3 minutes. Some television shows such *as Lost, Prison Break and 24* can be viewed as mobisodes.

The first mobisode was based on the popular television series *24*. The original television show is adapted into twenty-four one-minute mobisodes. The premise of the story centers on a character a CIA agent, Jack Bauer, who has twenty-four hours to save America from subversive terrorists. *Haunting Melissa* is a story created as a series of digital stories exclusively for the iPhone and iPad. With the viewer getting information given out in random and unanticipated ways, this mobisode uses the technology to the full extent and creates a new framework for digital storytelling. Designed specifically for the features of a smart mobile device, the mobisodes utilize the technology of these devices in a way that allows users to interact with them. Each mobisode changes and is never the same upon second viewing.

D. PODCASTING

Podcasts are a series of stories that use audio, video, or text to "broadcast" the narrative or non-fiction, which can be downloaded from the Internet onto a portable media device such as an iPod, mobile device, or personal computer. Podcasts can be produced by anyone, from professional podcasters such as NPR (National Public Radio)—favorites include *This American Life and Prairie Home Companion* both weekly radio shows—to amateurs. According to dictionary.reference.com, a podcast is a "digital audio or video file or recording, usually part of a themed series that can be downloaded from a Web site onto a media player or computer." End users can subscribe to a particular lecture or radio show or create their own podcast using an audio recording device. Many podcasts do not use video and are easy to create using iTunes.

E. THE ROLE OF SOCIAL MEDIA IN DIGITAL STORYTELLING

Each day social media sites such as blogs, facebook.com, instagram.com and twitter.com showcase digital stories. The stories range in duration, theme, and plot. Social media–based stories can be interactive. Digital storytelling is used in social media from personal to commercial usage in particular when "branding" a product. Social media is used to tell the story of the product and keep the customer informed. In this TEDx talk by Jim Jorstad, he provides insight on effective ways to use social media to tell a digital story.

The film *Julie & Julia* was based on a blog, a form of social media, which documented the author, Julie Powell, cooking all 524 recipes from Julia Child's first cookbook in 365 days. This true story was made into a Hollywood film, proving social media is not a fad but an important and respected vehicle for people to collaborate and share stories and comments around a common subject. Social media can promote a digital story to a larger audience, but in the end, the story has to have value.

Twitter.com offers the end user text-based digital stories using tweets, which are limited to 140 characters per tweet. First created in 2006 by Jack Dorsey, Evan Williams, Biz Stone, and Noah Glass, in 2012, there were 500 million tweets. Seattle Noir is a Twitter-based digital storytelling project composed of stories that combine noir fiction with life in Seattle, Washington.

Fig 4.3

Children dancing at a live performance

F. PERFORMANCE

A well-crafted narrative, whether traditional or digital, is fundamental to any story's success, whatever the format. This is true for live performances (Figure 4.3) or broadcast. Digital effects give performances an additional robust entertainment value. A digitally enhanced performance uses computerized effects to heighten the story. The Broadway shows, Phantom of the Opera, Wicked, and The Lion King, among others, all use digital technology such as audio and visual effects to enrich the audience experience. The NFL Super Bowl half-time shows combine live performance with digital effects. One of the most memorable was the Apple computer advertisement in 1984.

Using video conferencing technology, Skype, or Facetime, music performance has taken place with the musicians playing their instruments simultaneously from different locations around the world. Scott Deal, professor of Music Technology at Indiana University–Purdue University Indianapolis (IUPUI), directed a world premiere for the telematics opera "Auksalaq," performed by artists from IUPUI, the University of Virginia, University of Alaska, McGill University, Montreal, Norway, and other U.S. sites. Audiences were able to use their personal computers to watch and participate in the performance in real time.

G. CONVERGENT DEVICES AND DIGITAL STORIES

The method of connecting to the Internet changed with introduction of the PDA: personal digital assistant in 1997. This mobile device stored personal information such as contacts, calendar, data, and memos. Through technological advances, the PDA became more robust and evolved into a mobile device, a smaller, more sophisticated version of the PDA. The analog, pre-digital format was replaced with capabilities of easy access to text, media, and communication through the into one single device. [6] In 1997 Swiss Telecom Ericsson launched its GS 88 "Penelope" concept as the first smart phone. This mobile unit took on the name of "smart" because it had so much to offer the consumer. The smart phone such as the Apple iPhone has wireless capability, Bluetooth connectivity tools to send email, record video or still photographs, read, play games, watch videos, and listen to music. This all-in-one, Swiss army knife–type device has made the art of creating a digital storytelling fun and easy.

H. MEDIUM AND MEDIA

"Media" and "medium" are different forms of the same word. Medium is the singular form of media. The definition of medium is something that offers the user a platform to view or listen to a story (EBook, Mobile, Internet, games, performance, radio, television, film, and video). Media is the plural of medium and is used when a story uses more than one medium. With this, one Halloween night in 1938, Orson Welles, actor and director, read the H.G. Wells story, "War of the Worlds" to a radio audience over the Columbia Broadcast Network and sent a panic across America as some people thought the story was real. Because of the current ubiquitous communication system, a story such as "War of the Worlds" is unlikely to create the same reaction today.

Multimedia is developing content using more than one media such as: animation, text, graphics, photography, audio, film, video, time-based media and theatre production. Mass media is the communication of the story using a number of communication avenues such as the Internet and mobile phones.

An early animation and enduring media is stop motion animation first introduced by Ray Harryhausen. His stories include *Mighty Joe Young*, *Clash of the Titans*, and *The 7th Voyage of Sinbad* (1958). His first color animation was *Jason and the Argonauts* (1963). The process used by Harryhausen was tedious involving taking a single picture or frame, moving the characters and taking another picture until the story was complete. Using digital technology, stop motion animation can be produced by anyone owning a mobile device or digital camera. Director, producer and filmmaker Tim Burton has produced many digital stop motion animations including *The Corpse Bride*, *Frankenweenie* and *The Nightmare Before Christmas*.

Fig 4.4

Zombie illustration

I. The Story of "Superman" from Comic Book to Silver Screen

In 1938 two high school friends, Jerry Siegel and Joe Shuster conjured up Superman. This iconic character about a strong, handsome man who could fly came to be known as the "The Man of Steel." The character has continued throughout its 75-year history to protect the citizens and ensure justice in his adopted homeland. Each version retold the story in a unique way, using the cutting-edge media and technology of that time period. The story of Superman is an example of transmedia. In an article in *The Plain Dealer* in Cleveland, Ohio, Michael Sangiacomo wrote about Superman and the evolution of the superhero story from comics into television shows, movies to merchandise.

The popular television series *Walking Dead*, first produced in 2010, ran for five seasons on broadcast television. This popular television series became a webisode in 2013 to enable the audience who enjoyed the television version to continue to view the story after the television series ended. Walking Dead is a three-part web series that was developed to last 7–10 minutes. The story *Walking Dead,* an epidemic of apocalyptic proportions has swept the globe, causing the dead to rise and feed on the living. In a matter of months, society has crumbled: There is no government, no grocery stores, no mail delivery, no cable TV. Rick Grimes finds himself one of the few survivors in this terrifying future. A couple months ago, he went from a small-town cop who had never fired a shot and only saw one dead body into the world of zombies (Figure 4.4) fighting for his life. Separated from his family, he must now sort through all the death and confusion to try to find his wife and son. In a world ruled by the dead, he is forced to finally begin living. First published as a comic it joined the ranks of a popular television series and is also available as a game.

The game consists of five episodes, released between April and November 2012. Telltale has also published retail versions of the complete game. Currently the game is available for Xbox 360, PlayStation 3, Microsoft Windows, iOS, OS X PlayStation Vita, and Kindle Fire HDX. An Ouya version is planned for release in early 2014. The game takes place in the same fictional world as the comic, with events occurring shortly after the onset of the zombie apocalypse in Georgia. However, most of the characters are original to the game, which centers on university professor and convicted murderer Lee Everett, who helps to rescue and subsequently care for a young girl named Clementine. Rob Kirkman, comic book writer provided oversight for the game's story to ensure it corresponded to the tone of the comic, but allowed Telltale to handle the bulk of the developmental work and story specifics. Three characters from the original comic book series make in-game appearances: Hershel Greene, Shawn Greene, and Glenn Rhee.

I. TRANSMEDIA

I. Types of Transmedia

What is Transmedia? It is not a new term. The transformation of media from one format to another has a long history. Consider the 19th-century stories about Frankenstein's monster and Sherlock Holmes. These are stories that have been produced from books to video/films, television shows, theatrical productions, interactive games, and merchandise (Figure 4.6). Transmedia storytellers must be cognizant of which media/multimedia to use because each adds a specific and unique contribution to the narrative. Star Wars, when first introduced in 1977, captured the imagination of millions and went on to grow from a story into an industry. The story, film, book, comics, animations, games, toys, and merchandise made it a classic transmedia success story. Because the story was created in a digital format it can be expanded to some of the following types of productions:

Spin-off
Comic book
Book
Social media
Motion picture/video
Website
Webisodes
eBook

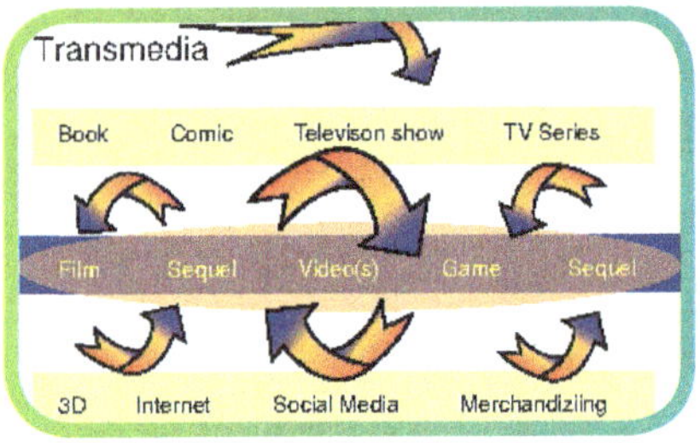

Fig 4.5

Transmedia

Game
Mobisode
App
Performance
QR code
Experiential opportunity/virtual world
Merchandise
Television
Film/video
Animation
Interactive

II. Definition

The term “transmedia” means a narrative synchronization across media. The collective term “media” (print, television, film, radio, video, games, computer and mobile) takes the content of the story and transports it from one medium, such as a book, to another, such as television series and a motion picture. From a production standpoint, it involves creating new experiences for the audience. Are you the type of person who reads the book first or goes to see the film on the silver screen and then reads the book? Consider the experience of reading the comic book Superman and then watching the animated version of that story in 3D. Keep in mind that it is important to choose the medium that will best convey the story so it won’t get lost in translation. The phrase “The medium is the message,” coined by Marshall McLuhan, is important to transmedia because the medium is the conduit of the message. The question is how effective is the medium in telling the story? Some stories translate well from one medium to another and some do not. Transmedia appeals to a wide audience and to industry because it extends the life of the story. The audience can influence the popularity of a story. An excellent example is the success of the Harry Potter, books turning into films, games and theme park, The purpose of writing a trans mediated story is that one medium is not enough to tell it all. According to media scholar Henry Jenkins [8] “the art of digital storytelling represents a process where integral elements of a fiction get *dispersed systematically across multiple delivery channels* for the purpose of creating *a unified and coordinated entertainment experience*. Ideally, each medium makes its own *unique contribution* to the unfolding of the story." [11]

Oftentimes, transmedia stories contain copious amounts of online elements in which users will share and talk and in the process driving "viral" marketing. Excessive marketing can possibly dilute the quality of the story. The purpose of a transmedia story is to supplement the core story. The importance of creating successful transmedia is that the creators of the supplemental media pay close attention to maintaining the continuity of the story. It's important when creating transmedia stories to recognize the signature contribution of that specific medium to the unfolding of the story.

III. Properties of Transmedia

The story must fit the individual media so that the audience has the best possible experience. Transmedia is often used to extend the life of the story and also can be developed to personalize the experience for the end user.

i. Audience engagement is critical to the success of transmedia. The quote from the movie Field of Dreams, "if you build it, he will come" is not always the case. The audience participation is vital and should be part of the initial concept and goal.
ii. Transmedia tips include understanding your audience and giving them a meaningful experience. Develop all parts simultaneously so they fit together, and make sure that each medium selected contributes something unique to the digital storytelling experience.
iii. Marketing options

Transmedia can be used for promotion and marketing of the digital story by expanding the story as a commodity, such as a novel that was turned into a motion picture and a toy available at fast food restaurants. Businesses can use this opportunity to entice end users to view or read the story and redeem prizes and attract a wider audience.

EXERCISES FOR STUDENTS

a) Discuss the similarities and differences between traditional and non-traditional stories.

b) Research and discuss stories that are interactive including stories for games.

c) Discuss convergent devices and digital stories.

d) Research examples of transmedia.

e) Develop a transmediated story.

f) Create a story for social media.

5. DEFINITION OF DIGITAL STORYTELLING AND HOW IT IS USED

Paraphrasing Andy Warhol (1928–1987), an innovative filmmaker and artist, every person will have fifteen minutes of fame in their lifetime. Warhol in his time believed that everyone has a story to tell. With the introduction of the camera-equipped cell phone, anyone can produce his or her story and share it through e-mail, social media, and websites. Stories are created for a variety of reasons that range from personal to professional. Digital storytelling takes the oral or written story or narrative and digitizes it through the use of technology. Digital storytellers use multimedia to enhance the narrative. This includes live action and animation, text, visuals, audio, and can also be interactive. One of the earliest films that combined live action and animation was "Anchors Aweigh" (1945) starring Gene Kelly and the cartoon characters Tom and Jerry. In the 1988 movie *Who Framed Roger Rabbit*, the actor played by Bob Hoskins partners up with "toons" to solve the crime. Digital storytelling is used in a variety of ways such as the following:

A. COMMUNITY ENGAGEMENT

Digital stories about members of a community from large to small are becoming more commonplace. Storycorps.org is an independent nonprofit organization that has been collecting stories since 2003. Over 45,000 interviews have been recorded and are in the collection of American Folk life Center at the Library of Congress. The digital story of a community can be a powerful framework to showcase individuals from that locale. A story about New York, a city of approximately nine million people, can be seen on a more personal scale through the project Humans of New York, an online site and social media network that was created for New Yorkers to voice their stories. This anthology has become a collection and reflection of the inhabitants of New York City. The important part of telling a community-based story is to give it a common theme. The Indianapolis Motor Speedway is archiving stories about the annual race in Indianapolis, Indiana. Amassing stories and collecting large amounts of data is futile unless the stories share subject matter. into a story worth telling.

B. CULTURE AND TRADITIONS

Digitizing the stories of a culture or tradition is a great way to preserve and archive these treasures. Stories passed down through oral delivery are part of the human experience. The digital generation has taken the oral story and digitally preserved these stories and traditions through video and audio recording devices. Stories told around the dinner table can be digitized and shared. The traditions and stories of a family's heritage can be a source for older members to reminisce and inspire younger members about their culture and heritage. Yumi Matsui, a Humanities/ English teacher at the Life Academy in Oakland, California, and consultant with the Bay Area Writing Project, worked with immigrant students to create digital stories. "The Digital Storytelling and the Common Core" focuses on the experience of researching, writing, and producing a digital story. Digital stories can be created using archival media such as old photographs, 8 and 16 mm film, and VHS tapes. Virtual tours supported on mobile devices are gaining popularity. A digital map can give the end user the ability to see important sites and artifacts without being on location. The 360-degree virtual tour of Rome takes the end user on a tour with music and impressive visuals. Another example is to tell the story of the local residents and put face to place for the end user who wants to know more about the cultural aspects of a location. The Digital Map of Paros is an example of capturing the stories of local residents and their traditions of Paros, a Cycladic Island in Greece (Figure 5.1).

Fig 5.1

Fire Jumping Festival, Aliki Paros, Greece

C. EDUCATION

Lessons that use multimedia—video, audio, motion graphics, animation, and/or interactivity—keep students more engaged. Creating digital stories for educational purposes provides students with a method to connect the past, present, and future. The digital story enhances student's ability to understand the subject material when it is presented with multimedia. An example is to have students listen to a story, write about the story they listened to, create a script of the story, and finally digitally record the students' interpretations of the story. National Public Radio (NPR) hosts an annual contest for storytellers to write a three-minute fictional story.

In the round of stories each year, judges select one story from thousands of entries. Listening to and/or watching stories in a digital format give students a better understanding of how to use technology as a learning tool. They learn to record, edit, and produce digital stories about any subject matter. There are many ways digital storytelling can be used in the classroom. Lessons that involve students using mobile devices for research and recording information enable students to create their own stories. Digital stories have the educational benefit of engagement and providing opportunities for students to be active learners and participate in the subject matter. Students enhance their skills in writing, media literacy, and technology by creating narratives. Digital storytelling provides an avenue for students with different learning styles (aural, visual, and kinetic) to express themselves creatively and intellectually. Digital storytelling teaches time management, problem solving, and collaboration, which are very important skills for students to learn. A podcast, for example, of historic events told as stories is an excellent method of delivering content to students. Giving students links to accompanying information is important to promote research. Educators use digital stories to have students investigate issues and engage in research through interactive, dynamic learning such as in gameplay. Writing a digital story fosters critical thinking and analytic skills. More than anything, creating a digital story for educational purposes gives the student a voice to express and share a particular viewpoint. Using digital stories in the classroom enables students to make a connection between what they learn in the classroom and what goes on outside of the classroom. Digital storytelling encourages creative thinking and opens up new ways of telling stories that are blended with research, intellectual thought, and novel ideas.

D. ENTERTAINMENT

Digital storytelling is a form of entertainment. Generations have enjoyed traditional stories through hearing, reading, or watching them. With the advancement of technology, stories that are entertaining can literally come to life by watching one in a 3D IMAX theatre or interacting through gameplay. Digital stories are used for entertainment and to relieve tension. Digital stories are a way to transport the mind into a place whose purpose is to inform, inspire, transform, and purely entertain. Watch this stop-motion animation, "Gulp," a digital story produced entirely with a

Nokia cell phone. The artist *Scott McCloud* creates digital comics that offer selections of frames within frames that draw the end user into his stories.

E. MEDICAL

Doctors and medical researchers use digital stories in a variety of ways. An example is an informational digital story created by patients to share their experience with other patients who are facing a similar medical condition. Digital stories are also used to teach the patient about their medical procedure or surgery and in particular for children to ease the anxiety associated with going to the doctor's office or hospital.

F. PROMOTION AND MARKETING

The business of promoting and marketing a product and service uses digital storytelling from websites to inform, and social media to connect with their audience. The marketing of the narrative in itself is an interesting story. The success of the Harry Potter story was shared through websites and social media. The Harry Potter Facebook site has a lot of participants: 66,498,822 likes and 531,237 talking about this.

G. SELF-REFLECTION

Self-reflection or video diaries are a type of digital storytelling that is created to document personal thoughts, activities, and experiences. In this type of storytelling, self-reflection can be a tool for assessment of learning outcomes for instructors. Utilizing students' self-reflective/video diaries is an effective tool for students and instructors to think about learning outcomes through a different perspective. Sit a student down in front of a camera equipped with video and audio capability and ask the student questions related to their experience in the course. Recording students can be used to assess students about which learning outcomes were most meaningful.

Students from Indiana University School of Informatics and Computing at IUPUI used self-reflection videos as testimonials to their experience in a study-abroad class, "Digitizing Cultural Heritage Paros, Greece." These students were asked to describe their understanding of the communication, culture, environment, language, and involvement with the residents of Paros once a week over a month-long study-abroad program (Figure 5.2). Give students the choice of writing a report, the traditional assessment method of lessons learned, and the self-reflection video option. Both have their advantages; however, as the saying goes, "A picture is worth a thousand words."

Fig 5.2

Study abroad students in Paros, Greece

H. THE ROLE OF SOCIAL MEDIA IN DIGITAL STORYTELLING

Social media began in 1969 when CompuServe used a dial-up method to connect two parties along a shared network. The AOL network became available in 1985 followed by MySpace and LinkedIn in 2003. The following year, Facebook and Flickr came on the scene. According to Forbes.com, the number of social media online sites has grown from 8% of U.S. adults in 2005 to 72% in 2013. In regards to social networking viability, the past has proven that a newer and better social network is always on the horizon. With each new flavor or version of social media, the purpose has always been to enhance the connectivity of people from all over the world. Blogs, Facebook, Google+, Podcasts, Pinterest, Instagram, LinkedIn, Reddit, Tumblr, Twitter, and Vine are some social media sites. From 2006 to April 2010, The United States Library of Congress made an agreement with Twitter to collect tweets that were part of the story of America project. This collection of approximately 170 billion tweets will become part of the Library of Congress archive and be available for research purposes. Social media has become the platform for readers to comment on, critique, and participate in a shared conversation. Each conversation is a digital story providing people around the world an opportunity to tell their story and comment about a common subject. Social media is used to complement stories, activities, and experiences. It is a digital tool to connect different media. Examples are television programs such as *American Idol* and *Dancing with the Stars*. These programs use social media to involve audience participation. Winners of these popular television shows are based on judges and votes from viewers using social media.

EXERCISES FOR STUDENTS

a) Discuss stories that showcase community engagement.
b) Research stories for education, medical, etc.
c) Research and discuss stories about culture and traditions.
d) Discuss the role of social media in digital storytelling and find examples.
e) Write a story for education or for medicinal purposes.
f) Discuss the role of social media in digital storytelling.
g) Create a self reflection story.

2 Part Two DEVELOPMENT PROCESS

6. RESEARCH

A. IMPORTANCE OF RESEARCH

Before you dive into creating a digital story, do yourself a favor and read, view, or listen to different types of stories. This will expose you to a greater understanding of what makes a good story. Most everyone has a favorite story. What is it about the story that appeals to you? The benchmark of a good digital story is that it makes a connection to the audience. The story has touched something that is important and meaningful to the end user. Digital stories need an engaging plot, believable characters, and appropriate multimedia to enrich the narrative.

Digital stories often use the same themes as traditional stories: friendship, life, death, love, war, revenge (Figure 6.1). Analyze your favorite story (movie, book, etc.). What are the elements that make it your favorite? What is the main theme? What is the main character's primary goal/problem? How do they overcome the obstacles standing in their way? Does the character change over time? These are part of the research that needs to happen before writing the story. In an online article, Christopher Booker explains the basic plots of storytelling. Research is also needed to understand your target audience. Who will be reading or viewing your story? It is important to know who your audience is and how they might react to the digital story.

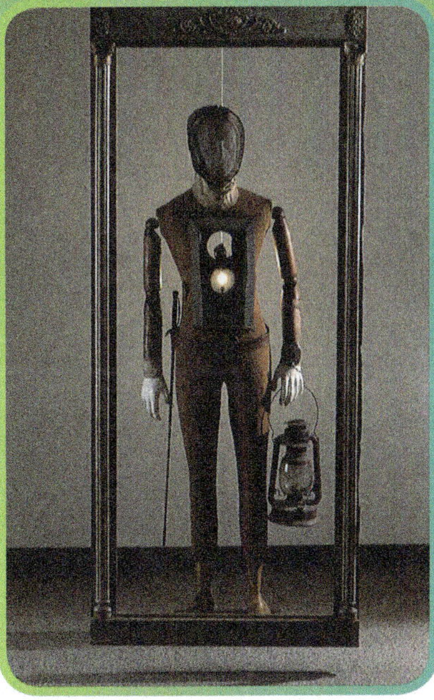

Fig 6.1

Looking Glass

Make a list of some good story plots that you are interested in reading or viewing. Ask your friends, colleagues, classmates, whoever, to recommend stories for you to read or view. Make notes of what is happening to the main actor(s) and in your research, answer the following questions about each story:

1. What is the plot of your story?
2. What are the cause and effect, the main complication, failure, or journey that the main actor must undertake to resolve the story?
3. Is there a clear protagonist and antagonist?
4. Does the main actor(s) succeed or fail in their final effort?
5. Are there subsequent efforts on the part of the main actor that resolve and achieve his/her goal?
6. What is the main actor's goal?
7. What is the relationship between the main actor and the co-actors?
8. What are the subplots and what is the main actor part?
9. Describe your actor's characteristics
10. Describe the world where the story takes place, the setting
11. Did you use Freytag's pyramid to plot out your story?

B. CHOOSING A THEME

The theme is the underlying message of the story. Theme-based stories have been around since Shakespeare. His themes are timeless. Comedy, drama, and romance have been portrayed in most every media. In the story *King Kong,* the theme is love between beauty and the beast. In Romeo and Juliet, it is love between a man and a woman. Love is Love. In the book *Jurassic Park* by Michael Crichton, the theme was cloning the DNA of dinosaurs and bringing them back to life.

In the story of Frankenstein by Mary Shelley (Figure 6.2), a monster is brought to life from dead body parts after being subjected to electric shocks. Both of these stories are based on the same theme: man versus science by pursuing dangerous knowledge. The *Zombie Story* (Figure 6.1) compares *The Walking Dead* with an animated feature film.

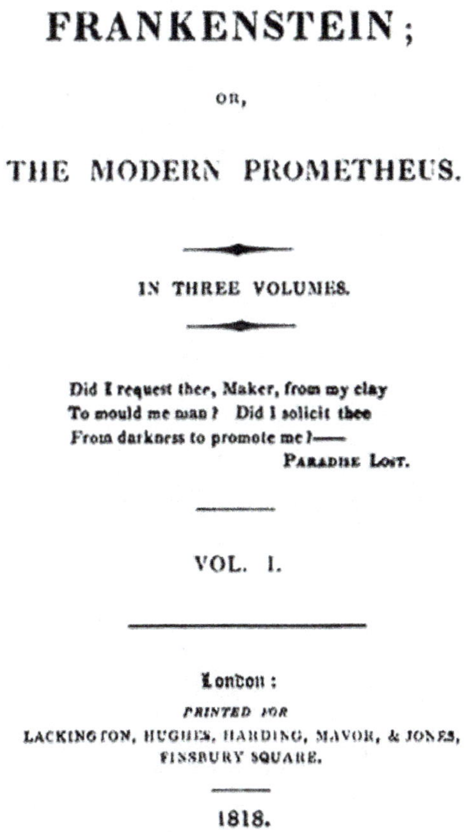

FRANKENSTEIN;

OR,

THE MODERN PROMETHEUS.

IN THREE VOLUMES.

Did I request thee, Maker, from my clay
To mould me man? Did I solicit thee
From darkness to promote me?——
PARADISE LOST.

VOL. I.

London:
PRINTED FOR
LACKINGTON, HUGHES, HARDING, MAVOR, & JONES,
FINSBURY SQUARE.

1818.

Fig 6.2

Original manuscript of Frankenstein by Mary Shelley

6.2 Mary Shelley, Frankenstein; or, The Modern Prometheus, Vol. 1, 1818. Copyright in the Public Domain.

Here are some themes to choose from when contemplating the main idea of your digital story:

- Man vs. Man (external struggle)
- Man vs. Himself (internal struggle)
- Man vs. Society (moral struggle)
- Man vs. Nature (struggle against fate)
- Man vs. Science (struggle with the unknown)

1. Love conquers all
2. Loss of innocence
3. Overcoming the monster
4. Hero must defeat a monster and restore order to a world threatened.
5. Rags to riches
6. Modest, generally virtuous but downtrodden characters, who achieve a happy ending when their special talents or true beauty is revealed to the world at large.
7. Good vs. evil: the battle
8. Revenge
9. The quest and the battle
10. Voyage and return
11. Triumph over adversity
12. Not always synonymous with humor. Instead, the plot of a comedy involves some kind of confusion that must be resolved before the hero and heroine can be united in love.
13. As a rule, the terrible consequences of human overreaching and egotism.
14. Death
15. Rebirth

    A threatening shadow that seems nearly victorious until a sequence of fortuitous (or even miraculous) events lead to redemption and rebirth = restoration of a happier world.

C. CHOOSING THE TYPE OF STORY

Writer and director *Woody Allen* is famous telling stories that combine different themes. His combination of drama and comedy in films like *Annie Hall, Manhattan,* and others make his style of storytelling unique. He said in a tweet message, "It's just storytelling and you tell it."

Listed below are some types of stories to consider. Be inventive and create a story that you think is worthy of telling.

1. Action
2. Adventure
3. Biography (Biopics)
4. Buddy Film
5. Comedy
6. Crime
7. Disaster Movies
8. Documentary
9. Drama
10. Epics
11. Fantasy
12. Film Noir (P. I.)
13. Gang Movies (Mob)
14. Horror
15. Romance
16. Martial Arts
17. Murder Mystery
18. Musical
19. Sci-Fi
20. Suspense
21. War
22. Western

D. AVATAR: BEHIND THE SCENES WITH JAMES CAMERON

The motion picture *Avatar* written and directed by James Cameron, is available in both 2D and 3D. The story is about a world of alien creatures that are attacked by human beings. This type of story, involving a conflict between human and aliens, and the exploitation and acquisition of the indigenous population for profit is a popular theme.

E. THINGS TO CONSIDER WHEN WRITING YOUR STORY

It is very important to know who your target audience is because the digital story needs to make them care. The characters in your story must be believable and engaging, and change over time. The audience must be able to follow the plot of the story and its characters. Storytelling is a journey and also a process. [3]

7. THE DIGITAL STORY STRUCTURE

A. PLANNING AND ORGANIZATION

Using this digital story structure will assist in your process of creating a digital story. So before you begin to write the story, consider telling it to someone in under 30 seconds, as in an elevator speech. In that short time frame you should be able to explain the plot, goal, and characters.

B. WRITING THE CONCEPT DOCUMENT

Creating the concept is the first step to building the digital story. Every story has a beginning, middle, and end, so keep that in mind when you write your concept. A concept should be the main idea of the story. A concept document is short, two to four pages, and states the purpose and premise of the story. A concept document describes the plot in what, where, who, when, and why terms, keeping in mind the connection of the narrative to the target audience. It is the storyteller's responsibility to keep the end user engaged and caring about the story. A source for stories is radio.lab.org. As you listen to the stories on the Public Radio Station, try to visualize the plot and write down some concepts that appeal to you. Keep in mind as you write your concept document that this is the main idea so make sure it is comprehensible and above all engaging.

A concept document needs to have a hook that captures the audience's attention. In a nutshell, your story needs the "Big Six" answers to the following questions:

1. Who are the main and supporting characters in the story, and are they believable?
2. Who is telling the story? What is the POV?
3. What is the plot? (events/sequence of actions)
4. When does the story take place? (year, season, day, etc.)
5. Where does the story take place? (setting)
6. Why? (motivation/conflict)

C. WRITING THE TREATMENT

A treatment is used for television, film, and video. The main purpose of a treatment is to sell your story to a producer or publisher. It is a four-to ten-page document that contains the concept, budget, crewmembers, cast members, and schedule. It is basically the summary of the work that is going to be produced. The treatment is used in promotional and fundraising campaigns to summarize the work for intended benefactors. Kickstarter.com is a site that people use to raise funding for television, film, and video productions. The purpose of a treatment is to create a blueprint of the film, television show, or video, so everyone who reads it can visualize the end product. A treatment is not a script. A script has dialogue and camera angles and movement. The only technical information in a treatment is what multimedia will be used. A treatment is a presentation, a sales tool that enlightens the prospective investors about the digital story without having to read 100-plus pages and spending two or more hours of their time. Some treatments are text-based documents and others are rich with visuals and audio references. The treatment often is referred to as a director's treatment because it details the decisions made over the course of the production. It is, the look and feel of the story, the sound, and so on. One example of a treatment is discussed in an interview by Terry Gross from National Public Radio with Joel and Ethan Coen about the remaking of the movie *True Grit*.

D. CREATING THE OUTLINE

Making an outline can help you organize what you want to write. This is rough plan for your story and can help make the process of writing much easier. Writing the concept document should give you the structure of the story. Use this document to list the actions and events in a sequential order.

Creating an outline for a digital story using the Freytag structure (Figure 2.7) can be as simple as using this framework. An example story outline for a three-act play:

Act I Scene I

- I. Introduction: Thesis/synopsis
- II. Body of story
  - What is the story/scene about?
  - Where does it take place?
  - What happens?
  - Who is in the story/scene?
  - Why? Motivation? Conflict?
  - When is it?
    - Supporting scenes

Act I Scene II: Climax of the story/Antithesis

- II. Body of work
  - What is the story/scene about?
  - Where does it take place?
  - What happens?
  - Who is in the story/scene?
  - Why? Motivation? Conflict?
  - When is it?
    - Supporting scenes

Act I Scene III

III. Body of story: Conclusion/synthesis

- What is the story/scene about?
- Where does it take place?
- What happens?
- Who is in the story/scene?
- Why? Motivation? Conflict?
- When is it?
  - Supporting scenes

E. WRITING THE DIGITAL STORY

All good writers have one thing in common: they make their audience care. Stories are as old as human history and the framework hasn't changed much. Gathering around the campfire or dining table, the story has been a method of communication that enlightens and empowers the human mind and spirit. Writing a traditional or digital story is a craft. The skills you need are already in your toolbox. Begin by working your way through the digital story structure. Write stories about what you know, saw, or experienced. Telling stories about real-life events has an extraordinary effect on an audience. Look for inspiration around you through observations of life, from mall walkers to your friends and family. Go out and observe. Write in public. The great thing about writing a digital story is it can be done anywhere through the use of portable/mobile devices. Digital technologies are reshaping the power of storytelling.

No time in history has storytelling been so ubiquitous, so anyone anywhere can share a story with others connected to the Internet. Self-publishing is as easy as posting a story on any social media site. The question is whether or not the story is worth the reader's/viewer's time. Keep in mind as you write that how your story is told is as important as the story itself. "Does the dialogue flow naturally? Is there a logical and consistent flow of events? Are the events of the plot resolved in a way that makes sense in the context of the story?" The plot needs to be about a story that touches on something that is timeless: love, friendship, fear, etc. The theme is the controlling idea, the central insight of what the author is trying to convey. The characters need to be believable. Time and place constitute the location or setting of the story. Style, tone, and pacing add depth to the story. These elements are part of unraveling of the story that give it dimension. The point of view (POV) refers to how the story is being told. From the writer's point of view, first-person perspective, the author tells the story as one of the characters: "I." Third-person narrative tells the story from the vantage point of an outsider looking in on the story. This can be portrayed as knowledge of a specific character telling the story of another character. This is a very typical point of view for fictional narratives. The question of what makes a good story can be answered by the audience's desire to want to keep watching or reading because they want to know how it ends. The digital story can become an experience by using digital multimedia: audio, animation, motion graphics, stereo graphics, video, and/or interactivity. A principle of good digital storytelling is to give the audience just enough information at specific juncture points of what they need to know when they need to know it. The intention is to have the audience engaged enough to want to know more. The hard part of creating a digital story is not being formulaic.

Here is a story idea: A character boards an airplane and settles into their seat. Right before takeoff, the character glances to the person sitting next to him/her and realizes that when they were children, this same person was their worst enemy. What happens next is what makes this story soar, or sink. A method to know if you are on the right track to writing a "good" digital story is to write it without using multimedia and technology. Write the story using the traditional three-act play structure and then revisit the story and embellish it with multimedia. The story will dictate which multimedia is appropriate to use. The process is to write the story and then shift the lens of the narrative to utilize the most appropriate multimedia to make the story come alive and be memorable to the audience.

Fig 7.1

Gertie the Dinosaur by Windsor McCay

Star Wars would never have achieved the cinematic artistry without the inventive accomplishments of Industrial Light and Magic. According to Sparknotes.com, "practically every frame of a *Star Wars* film has some sort of effect added in." From the beginning of time, storytelling has been a craft. In fact, special effects is not new, it have been a part of storytelling since the early 1900's. One of the first innovations was in animation. In

1906 *Humorous Phases of Funny Faces* because the first animated short and in 1914, Winsor McCay's *Gertie the Dinosaur* (Figure 7.1) became the first animated cartoon. The first full-length animated film was Walt Disney's *Snow White and the Seven Dwarfs*. As we have learned, stories that were written in a traditional format can be converted. Digitizing a story gives the story longevity; archival treatment allows it to be viewed, read, or experienced for generations to come.

F. DIGITAL STORY STRUCTURE

I. Definition

Many digital stories have complex plots that involve subplots. This is an ideal environment in which to create interactive stories. Zombie vs. Humans are non-traditional interactive story-based games. First presented in 2005 at Goucher College, in Towson, Maryland, it is now played at over 650 colleges and universities across the world.

Other films to review with non-traditional plots are:

1. *The Prestige* by Christopher and Jonathon Nolan and film
2. *Minority Report* by Philip K. Dick and film
3. *No Country For Old Men* by Cormac McCarthy and film
4. *Run Lola Run* by Tom Tykwer and film
5. Inception
6. Sixth Sense

A story using non-traditional plot structure is the story of the invention of the iPhone's voice information center, Siri. The film, Her, directed by Spike Jones, is not the conventional love story. It's about a lonely writer who develops a relationship with the voice command from his cell phone. Another example of creating a story by using digital technology is Apple's *Misunderstood*. *The Broadway production of The Lion King* used digital technology (lighting effects and sound) with shadows and puppetry to bring a child's tale into a 3D space filled with adventure and drama. The television program *Twin Peaks* was an idiosyncratic story that used flashbacks and foreshadowing about an FBI agent investigating a murder.

Using Freytag's traditional story structure doesn't compromise the story. A three-act play remains a three-act play even if it is bathed in digital effects and multimedia. The *Star Wars* trilogy is a blend of myth and digital technology on a grand scale with prequels and sequels each based on Freytag's three-act structure (exposition, climax, and resolution).

Act I: Exposition

The beginning or thesis of the story must grab the attention of the audience. The characters are introduced at this point and the situation, conflict and goals of the narrative are established.

When writing this part of the digital story, consider the following:

- Introduce the situation, problem, goal
- Introduce believable characters (main and supporting)
- Introduce setting, when, and where
- Establish task, intention, desired outcome
- Establish obstacle, problem, adversary
- Create conflict, suspense, tension
- Weave subplots
- Usually ends with a reversal/setback/crisis (small win for antagonist)

Act 2: Climax

The Middle or antithesis highlights the conflicts. This is where the story reaches the highest emotional level in the story arc (Figure 7.4). The middle act (which is usually twice as long as Acts 1 and 3) develops and explores conflict, relationships using action and events (whether large or small) to move the story forward. This is where the catharsis occurs, especially after high drama is performed. Time to give the audience some time to breathe. There is a rise in action, inciting an incident that reaches a climax as either a compounding conflict/altercation or a comic relief/tension release.

When writing this part of the digital story, consider the following:

- Complicate the predicament/raise the stakes
- Introduce subplot

- Introduce subordinate characters
- Create overwhelming need for resolution
- Usually ends with a higher level of reversal/setback/crisis

Act 3: Resolution

This is the synthesis, conclusion, end of the story in which the plot is resolved. The character reaches his or her goal consequently bringing the narrative to an end. The final act is the denouement of Act 1 and Act 2, paying off all the development, strategizing and struggle that went on throughout the story.

When writing this part of the digital story, consider the following:

- Obligatory Moment
- Climax
- Deus Ex Machina—Divine Intervention
- Denouement/Falling Action
- Intensify the problem
- Resolve the subplot into the main plot
- Create an ultimate reversal/setback in protagonist's predicament
- Bring about resolution/denouement of final setback/story
- Create triumph of the protagonist/hero and downfall of antagonist/ villain

II. Conflict and Tension

Stories use cause and effect to set up conflict and tension in the plot.

- Conflict is required to drive the story forward
- Linear stories
  - Tend to make a boring story
- Extreme differences in power can also be boring
  - What fun is a "boss fight" when its hardcoded for the secretary to lose?

If you are having trouble deciding on a plot, try brainstorming. Suppose you have a protagonist whose husband comes home one day and say he doesn't love her anymore and he is leaving. What are actions that can result from this situation?

- She becomes a workaholic.
- Their children are unhappy.
- Their children want to live with their dad.
- She moves to another city.
- She gets a new job.
- They sell the house.
- She meets a doctor and falls in love.
- He comes back and she accepts him.
- He comes back and she doesn't accept him.
- She moves in with her parents
- She splits custody of the children with ex-husband.
- She takes the children and moves to France

The **next step** is to select one action from the list and brainstorm what could happen in the chosen scenario Conflict naturally leads to tension. Tension comes from *not knowing* what will happen next. In a love story for example:

- What will he/she do?
- How will they escape that pit/swamp/deserted island/etc.?
- Will their love survive?
- If they break up, will they get back together again?

- **Suspense.** Writer teases reader/viewer. Give the reader/viewer just enough at just the right time and don't give everything away.
- **Ownership.** Give both sides things to consider.
- **Advancement of plot.** Keep the story moving forward while intensifying the conflicts the protagonist faces.

- **Causality.** Hold fictional characters more accountable than you do real people. Characters who make mistakes frequently pay, and, at least in fiction, commendable folks often reap rewards.
- **Mishap/Misadventure.** Provide intrigue and keep the audience guessing so they don't lose interest.
- **Recognition.** Have the audience believe in the character(s). Have empathy for them.
- **Understanding.** Use a universal theme that the viewer/reader can identify with. Present a conflict that has universal appeal.
- **Summation.** Convince the audience that the ending meets the goal of the plot and resolves the conflict.

Tension "hacks" our brains ... we have a built-in desire to *know*, to find the solution. This is why spoilers are so bad ... a story without tension has no new information for us to acquire, let alone keep our attention.

Tension typically builds throughout the story and the protagonist will have to work through a number of situations in order to achieve his/her goal. Each situation will raise the stakes ... introduce new factors that may prevent (in building blocks) the protagonist from achieving his/her final goal. Ever-rising success or ever-rising tension can get old, so the story needs to have a hook of intrigue and hope to keep the audience's attention. Jeopardy is a type of tension. There might be life-threatening jeopardy that involves chase or fight scenes. There are unknowns in this scenario and questions such as "who did what?" Is someone misrepresenting the facts? What is archetypical of the part played by Tom Hanks as Robert Langdon in the Dan Brown thriller, *The da Vinci Code*? Consider in this story how much danger the main character(s) are in? Is it sharks, quicksand, an evil organization, and what will Langdon gain or lose in his pursuit of the truth? Adding in reversals of fortune/going from good to bad, the battle lost, love forlorn can keep the audience both interested and asking, "What's next?" This goes for protagonists and antagonists alike.

Aristotle said that reversal of fortunes, or peripeteia, is the most powerful part of a plot in a tragedy along with discovery. There is often no element like peripeteia; it can bring forth or result in terror or mercy, and in comedies it can bring a smile or it can bring forth tears. Building the conflict brings the most interesting and exciting aspects of the story because the audience realizes that soon the story will end. The writer Jane Burroway states that the crisis "must always be presented as a scene. It is 'the moment' the reader has been waiting for. In Cinderella's case, 'the payoff is when the slipper fits.'"

III. Hero's Journey

The *Star Wars* trilogy by George Lucas discusses *The Hero's Journey* (Figure 7.2) in the interview with Joseph Campbell. In the *Star Wars* saga there is intrigue, computer-generated characters, and digital effects.

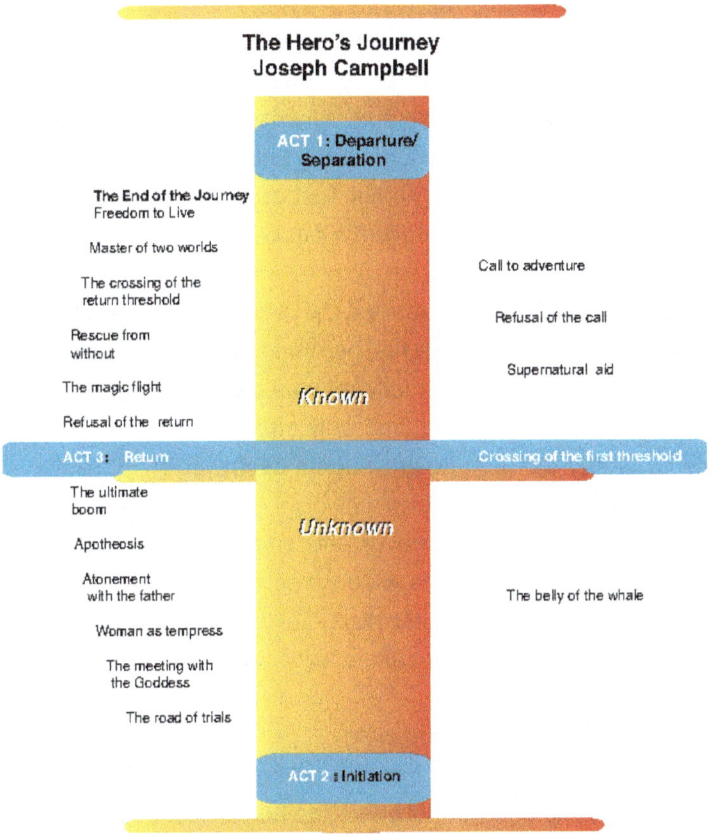

Fig 7.2

Joseph Campbell's *The Hero's Journey*

There is foreshadowing, giving the audience a sampling of events that will unfold as the story is told and subplots within subplots. The hero is a personification of a culture's mythology, traditionally described as:

- A warrior—the ideal of strength and courage
- An explorer—the founder of civilizations
- A philosopher—an explorer of the mind
- Artist and scientist—Campbell added these for the modern day

ACT 1:

- Departure (or Separation)
  The Hero feels restless (something is missing in life): a feeling of destiny.
- The Call to Adventure
  The journey starts with the hero in a conflicted state. The call to adventure is often told to the hero by a messenger or herald such as the Robot R2D2. In *Star Wars*, the hero, Luke Skywalker, begins the story frustrated about living on his Aunt and Uncle's farm because he wants to be a Jedi knight like his father.
- Refusal of the Call
  In some heroes' journey narratives, the hero refuses his/her call to adventure and then changes his/her mind and accepts the quest. In many stories, the hero initially refuses the call to adventure. In *Star Wars*, Luke leaves his home only after he finds his aunt and uncle dead and the farm he grew up on destroyed.
- Supernatural Aid
  During the hero's journey, he or she comes into a relationship with a helper. This can be a wise person who will give the hero advice and assistance. In *Star Wars*, Luke meets the Jedi Master Obi-Wan Kenobi, who gives Luke a light saber and instructs him about the Force.
- The Crossing of the First Threshold
  The hero finally must cross into the unknown. This can be a dark place such as the underworld where the hero will encounter the unknown, perhaps evil. Before the hero reaches higher ground towards

true enlightenment, the hero must cross the threshold between the known world and the unknown world. Often this involves facing the threshold's antagonistic guardian. In *Star Wars*, the threshold is Mos Eisley, a spaceport on the planet Tatooine, which acts as a doorway between Luke's home planet and the wider universe; Luke must avoid capture by the threshold guardians, the imperial storm-troopers.

- The Belly of the Whale
  The hero has defeated the threshold's antagonistic guardian and is now in a place of unknown reality. This is usually a dark place where the hero is tested to see how he or she can discover their true purpose, goals, and meaning behind the journey. The "belly of the whale" refers to perhaps an ambiguous place of dream-like forms. The name for this stage of the monomyth is based upon the story of Jonah and the whale. In *Star Wars*, the monomyth is the Death Star, in which Luke is engulfed and in which he learns how to be a hero.

ACT 2: Initiation

- The Road of Trials
  During his adventure in the underworld, the hero is continually challenged with emotional and physical obstacles that he or she must overcome. Many times this takes the form of a test, by which the hero sharpens his skills and proves his worth. In *The Empire Strikes Back*, Luke undergoes his training with Yoda.
- The Meeting with the Goddess
  After overcoming the road of trials, the hero often meets a goddess-like woman: beautiful, queen-like, or motherly. The hero faces the goddess and in doing so, faces another conflict. Campbell cites the lure of the woman as leading the hero astray. In *Star Wars*, the goddess to Luke is Princess Leia who wears white and appears like a vision initially.
- Woman as Temptress
  In some heroes' quests, the hero will encounter the goddess, but before he can unite with her, he must prove his worthiness by overcoming the temptation of her as a temptress. Sometimes

this stage is often lumped into that of the road of trials by many scholars because the hero, if he doesn't resist temptation, will be plunged back into the belly of the whale. In *Star Wars*, there is conflict between Han Solo and Luke because they are both in love with Princess Leia. This is resolved in *Episode VI—Return of the Jedi* when Luke discovers that Leia is his sister.

- Atonement with the Father
  The hero may encounter a father-like figure of patriarchal authority. "Father" and "son" are often pitted against each other for mastery of the universe. To understand the father, and ultimately himself, the hero must reconcile with this ultimate authority figure. In *The Empire Strikes Back*, Luke confronts Darth Vader and learns that Vader is his father; in *Return of the Jedi*, he is reconciled with the reformed Vader.
- Apotheosis
  The hero's ego is disturbed through his or her expansion of self-revelation. Often his/her idea of reality is changed, he/she may find him/herself able to do new things or able to understand things previously not known. In *The Empire Strikes Back*, Luke sacrifices himself rather than turn to the dark side.
- The Ultimate Boon
  The trials have almost ended for the hero and the goal is in sight.
  In *Star Wars*, Luke finally trusts in the force, manages to drop the bomb in the right place, and so destroys the Death Star.

ACT 3: Return

- Refusal of the Return
  Having found bliss and enlightenment in the underworld, the hero may not want to return with the boon. In *Star Wars*, Luke stays with the rebels rather than going back to Tattooine.
- The Magic Flight
  A mad dash is made by the hero to return with the prize. In *Star Wars* it is the assault on the Death Star.
- Rescue from Without

The hero may need to be rescued from without by humanity. In *Star Wars*, Luke confronts his father Darth Vader.

- The Crossing of the Return (third) Threshold
  Before the hero can return to the real world, he must confront another threshold guardian. The first threshold was a symbolic death; this is now a symbolic rebirth.In *Return of the Jedi*, Luke again confronts Darth Vader.
- Master of Two Worlds
  Once the final threshold is crossed, the hero is free to move between the two worlds at will. He has mastered the conflicting psychological forces of the mind. In *Return of the Jedi*, Luke becomes a Jedi.
- Freedom to Live
  With the journey now complete, the hero has found true freedom, and can turn his efforts to helping or teaching others in conflict. This is the end of the story, goal is reached, mission accomplished. In *Star Wars*, Luke becomes one with the Force and later is able to train others.

IV. Character Beats of a Story

Stories are structured so that tension between characters rise and fall during the three act structure. These peaks are called character beats, as they resemble an EKG diagram (Figure 7.3).

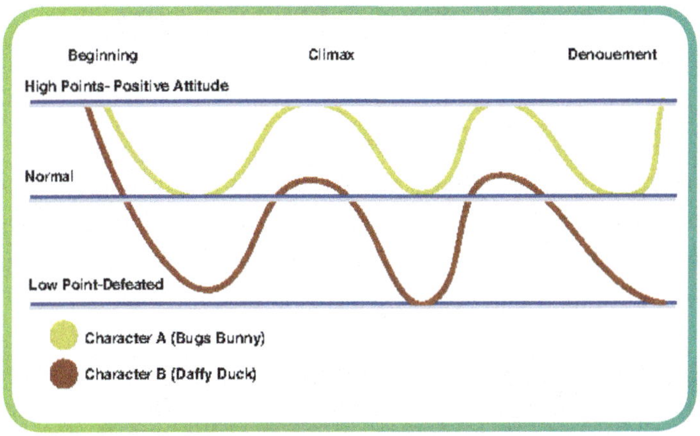

Fig 7.3

Character beats of a story

Some of the beats might be the character falling into a pit from which he/she escapes, but with a broken leg. It's the highs and lows of the story. Beats add tension and then release. The tension is still heightened as the character deals with new situations or conflict. Well-developed characters express a range of emotions. In this animation, *Rabbit Season Duck Season Trilogy*, Bugs Bunny's beats are mostly high when he and Daffy Duck try to convince Elmer Fudd which hunting season it really is, duck or rabbit? Bugs Bunny (Character A) remains confident-positive whereas Daffy Duck (Character B) encounters extreme high and low points, which makes their interchange funny (Figure 7.3).

To create a character beat chart using x axis (vertical) draw three lines on the X axis: bottom is low depressed, fails or otherwise is the bottom of the character(s) worst case scenario; middle line represents

Fig 7.4

Story arc

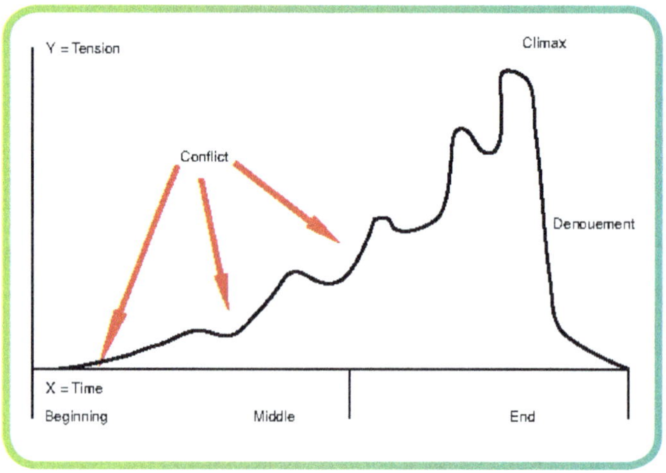

normal behavior, everything is okay and the top line the character(s) are succeeding, feeling great outwitted the antagonist. The y line (horizontal has three points on the line: at the x/y juncture is act 1, mid way act 2 and end act 3. Give your character(s) a mark (x or o) and mark the highs, normal and low points of your characters behavior or actions during the scene.

V. Story Arc

A story arc is a method of charting the narrative's conflict and tension. This is true in most action-adventure types of stories. It's important for the audience to get a general rise of tension from a compelling story that has subplots within subplots but also to reward them with a more comforting scenario such as a love interest or raising a child. Read any novel by Tom Clancy and take a ride into espionage and secrecy. Such Alfred Hitchcock films as *North by Northwest* and *Psycho* are intricately woven stories with romance and drama, tension and release. Examine the times where there is conflict and how story progresses with low and high points until it reaches the conclusion (Figure 7.4).

VI. Continuity, Timing, and Pacing

Timing and pacing refer to the speed at which the story unfolds. Utilize the story arc to determine the highs and lows of the story in collaboration with character beats. Consider when in the story the pace should quicken and when it should slow down. Give your audience breathing room, (catharsis), time to contemplate or figure out the plot. You will learn after writing a few stories how to use timing and pacing to intensify the action at just the right juncture. The dialogue of the story will also play a factor in how fast the story unfolds. Speaking slowly conveys the idea of introspection, disbelief, or awe. Speaking too fast can lose an audience; too slow and people get bored. Pacing is about finding the right tempo to keep your story moving forward. Variation is pacing at the right times to intensify the scene and to slow things down. Using the story arc and beats of a story will assist you in knowing how to time differently paced scenes so the story unfolds perfectly.

Continuity, timing, and pacing of the digital story will differ from traditional stories if interactivity is part of the scenario. In games, for example, the player(s) determine the timing and pacing of their actions. In film, television, and video, much of the story is cut to accommodate the time factor if viewed on a broadcast network. Get rid of the fat, so to speak, in an effort to move the story forward. Using digital technology, the assembly of shots uses purposeful cuts and montage techniques to build the story in a meaningful way and also to keep the attention of the end user. Many films are shot out of sequence so the pacing, and timing of the narrative are very much connected to editing in post-production. Continuity is important to maintain, so the parts that are edited out are unnecessary to the story. For example, if an action film takes five minutes in real time, the viewer will most likely get bored. Cutting out all the monotonous parts and keeping only that which keeps the continuity of the story moving forward will keep the reader's/viewer's attention. Giving the reader/viewer an opportunity to fill in the missing parts by using their imaginations and allowing for breathing room have all the characteristics of a well-timed story. In the Bugs Bunny cartoon *Conducting an Opera*; notice the change in pacing in characters' voice and body movements as the music escalates. A similar scene is at the end of the movie *Bye Bye Birdie*, when the conductor speeds up the ballet and changes the tempo thereby creating a very funny scene.

VII. Flashbacks and Foreshadowing

A flashback creates a series of sub-stories that follow one after the other and come together at the point of resolution. This type of story can have multiple flash backs, flash-forwards, sub-plots, and antagonists and protagonists. In this type of story, the plan and organization are essential for success. A modification of the traditional three-act play with each act having a series of inner acts is similar to Russian dolls that fit inside one another. What you want to avoid is a story that is so involved it is confusing to the audience. Similar to entering a haunted house, you don't know what is behind each door, so you proceed slowly and each door you enter adds to the plot. The writer and director Robert Altman used flashbacks in many of his stories and films. His approach to using flashbacks was to construct a past story with the present like concentric circles (each circle a unique story in a different time frame). This approach connects the stories at one special scene, and jumps between the past and present stories at specific points along the storyline. The non-traditional story usually follows the first act and then things change in the start of the second act or at the denouement in the third act.

Foreshadowing refers to something in the story that has not happened. John Steinbeck used foreshadowing in his classic novel, Of Mice and Men. In the story one of the main characters, Lennie, kills a mouse early in the story foreshadowing his ability later on to kill a woman.

G. CHARACTER DEVELOPMENT FOR DIGITAL STORYTELLING

Characters for games, animation, and live action take a different route because they require different treatment than a traditional story. The lesson to learn here is what makes a good character. What is the relationship of the character(s) to the plot? What is the role of the character in the story? What are the best supporting roles and what is the relevancy between the lead actors and the supporting or minor ones? Your story character(s) needs to be memorable and it is your job as a writer to create with that objective. A well-developed story needs to have the characters change over time. The bottom line to writing character(s) is to make your audience care about them.

Fig 7.5

Character archetypes by pengWu

I. Character Archetypes

Character archetypes are a great way to understand what drives a story (Figure 7.5). Consider all the main characters in *Star Wars, Return of the Jedi* and see how they might fit into these traditional archetypes:

1. Protagonist: the hero of the story—Luke
2. Antagonist: the villain of the story—Emperor
3. Mentor: the advisor to the main character—Yoda
4. Love interest—Princess Leia
5. Supporting character—someone who helps the hero: Han Solo, R2-D2, C-3PO

When choosing an archetype, consider the framework and also the motivation. "My character is a fisherman." Define what makes him a good fisherman. Can he drive a boat, fish, and survive on the water? Define how he/she became a fisherman. Does he/she have an inner desire to live near water and fish? Was he influenced by Melville's Moby Dick (downloadable free from Project Gutenberg), Hemingway's *Old Man and*

Fig 7.6

Man walking with chair and guitar

the Sea, or Captain Jack Sparrow's character in *Pirates of the Caribbean: The Curse of the Black Pearl?* Was he brought up without water to fish or swim in? Decide what being a fisherman means to her/him. Now take a story about a businessman who encounters the fisherman. The businessman wants to purchase the fisherman's boat. What kind of framework is the businessman working in? The only way to develop a well-rounded character is to write about the character as if he, she, or it was a part of your life. Develop character biographies by investing time in working out the idiosyncrasies, backstories/flashbacks/foreshadowing and psychological profiles of each of your characters, main and minor. What are their personal and professional histories? What are their hopes and fears?

Note how much detail is applied to the character as you craft the history and present of a character: they begin to "come alive" in your mind. What is the story about the man holding a guitar and chair? Where is he going and why? Who is he? A good exercise is to create a character biography from a photograph such as this one (Figure 7.6). What are your character's physical and emotional attributes? If you take time to develop your character's idiosyncrasies and unique characteristics it will be easier to predict their actions and changes throughout the plot. Here is a sample blank bio. Give your character(s) standout personalities with strong names and distinctive features. In video game stories, character names are very important, including Duke Nukem, Max Payne, Earthworm Jim, and of course Sonic the Hedgehog. Film character names include Indiana Jones and Luke Skywalker.

II. Game and Virtual World Character Development

In video game character development, control or agency is very important in the design process. A character is a representation of the player and is commonly referred to as an avatar. The player in an ARG (alternative reality game) might have a lot of power or be limited and have less agency/control over what they look like and what actions they have. Users can choose characters in games and *virtual worlds*—such as *Second Life, Skyrim* and *World of Warcraft*—dress them, give them roles and responsibilities, and associate the with or play them against other end users. The stages of character development for an animation, film, game, or stage performance begins with a rough character sketch that embodies the appearance and personality of the character.

The most important attribute of a character's personality is to develop her or his weakness and their goal in the story. Through the development of the character's personality, the writer needs to understand, "What is that personal problem inside that is hurting the hero?" Solve the problem by going after a goal. Plot comes from the character(s) you create and a goal for the main character that will force him or her to deal with their strengths and weakness. The plot comes from this development, how characters overcome their weakness and gain answers; do this properly and you will have told a great story.

Archetypes have been a staple for character development and no two are alike. Take this type of character who arguably was the main character for Thomas Wolfe's character Sherman McCoy, played by Tom Hanks in the film *Bonfire of the Vanities* and Michael Douglas as Gordon Gekko, in *Wall Street.* This character type decides to use his/her powers for money and is loyal to whoever pays him/her the most. Here is a list of information about personality types:

- Myers Briggs
- Enneagram
- Alignment

III. Building a Character Biography

Building a character biography for a story is a process well worth your time and effort. After completing the concept document, a writer should have more insight into his or her character. The first step in character building is to create a description of the character Here are some things to consider as you build your character biographies:

- Name
- Age: description/appearance
- Dreams/goals/ambitions

- Weaknesses/strengths
  - Greatest flaw and best quality
- Past
- Family relationships

One method to develop the character is to interview him, her, or it. Decide what questions would be appropriate for the character as if he or she were an interviewee for a talk show on radio or television. Or better yet, act out the character and have someone ask the person questions as if he or she is the actual character. If your main character asked you to lunch, where would you go and what would you order to eat? What would you talk about? This will help you when you write the dialogue for the digital story. Another method of building a character biography is to have each of the main characters write each other a letter describing themselves and their goals. For example, have the protagonist write a letter to the antagonist and vice versa. If you are writing an action-adventure type story, have the main character describe what he or she does during their down time when they are not chasing the bad guys. This could be a very interesting activity to do with video game characters.

Character biographies are an interesting study, drawing the eye through the emotions of the actor and the mind of the writer. The changes that happen to a character and how they evolve and respond to situations are a core part of what makes the character interesting in the first place. Think about how a Ghostbuster would respond to a cyborg attack versus Super Mario. Consider in an action adventure story, in the aftermath, what would the respective characters' takeaways be? What are the rewards of their efforts?

- In *Tootsie*, Dustin Hoffman's character begins as a misogynistic chauvinist, but when he is forced to play the part of a woman, he also experiences a change in how he views women and becomes a different character by the end.

- In *Empire of the Sun*, Jim begins as a carefree young boy. After the Japanese take over Shanghai and he is separated from his family, he is forced to suffer trauma because of the war.
- In *The Godfather*, Michael Corleone at first does not want to have anything to do with his father's crime business. When his father is attacked and barely survives, Michael begins a war of retribution against those responsible. This effectively and ironically sets him down the path to becoming the head of his father's crime syndicate. *The Godfather Part II* chronicles Michael's effective fall from grace as a result of becoming a powerful crime lord.
- In *Taxi Driver*, Travis Bickle degenerates from a somewhat disturbed, highly disorganized Vietnam War veteran into an extremely well organized, full-blown psychotic.

Developing a character biography is similar to developing character beats (Figure 7.3). Lay out your characters change over time by creating a graph to mark changes. The character biography may extend over to the next story, a sequel, or another episode and into the transmedia arena. In episodic TV series, the character bios have become a hook that writers often use to ensure viewers continue watching.

Plot out the changes—the emotional, physical, and psychological transformation(s) of the character(s) in the digital story. Use the story arc (Figure 7.5) to pin point the scene(s) in which these changes of character occur. A character bio includes the place in the story that the change in the character occurs as well as how he or she changes. The change of character is ultimately what makes them interesting such as how he/she goes from positive attitude to defeated? Change can be also something learned, as in mastery of a new skill. Consider how the character changes during the evolution of the story. How does Luke change to overcome the Empire? He develops Jedi skill, learns to be a team leader, and decides through conflict not to fight the Emperor or kill his father. In the video game *Grand Theft Auto* (GTA), the main character change as he works his way up the game world. The change from newbie to trusted associate is his character bio's growth. Very common in multiplayer online games, such as depicted in this PSO (Phantasy Star

Online game) is the ability to customize the sex, features, and clothing of the characters.

IV. Motivation

Motivate your characters: your protagonist, antagonist, and other important characters and give them clear goals. This will energize them and make them understandable. What are they striving for and why is it so important to them? What are the stakes? What will/could/should they win/lose?

Make the characters vivid. Interactive media, as opposed to film, does not leave any room for subtlety. Your characters' most important traits should shine through clearly. For models, study comic strips, graphic novels, films, and animations, all of which use different techniques to develop a character.

The protagonist is the character you follow, perhaps the character you root for. This character is not always "good." In the Batman and Spider-Man stories, these two superheroes have to prove their worth to the residents of their hometowns. Both start out as anti-heroes and typically win the hearts of the locals and audience. The protagonist needs to work towards a goal and overcome obstacles. The antagonist puts up obstacles that the protagonist must overcome. Develop all parts of the character from their appearance that will give your reader/viewer a visual understanding about the character.

Avoid stereotypes at all costs. Stereotypes can make your work seem predictable and bland. Either avoid them completely or take a familiar stereotype and give it a twist. For example: make your mad scientist character a little old lady, or have your tough enemy soldier be an expert on French cuisine. Give characters a distinctive look. The appearance of your characters can telegraph a great deal about them. Just as important as appearance is the character's body language, how a character holds themselves, their walk-move (arms clenched to their bodies, stooped over a walker, a cocky swagger). Their clothing and object(s) they have on them, props, and so on are all details to creating memorable characters. In the film *The Big Lebowski* by Joel and Ethan Coen, the characteristic for "The Dude," was his appearance. He wore a bathrobe throughout the entire film.

The actions of the characters will show the reader/viewer what types of characteristics the characters have. This can be appearance and type of moral character, emotional intention, or habits. Make your character(s) believable and well rounded. Give the audience someone to identify with or absolutely loathe. Make the character fit the type of the story. For a horror story, make the main character scary; comedy, make them character funny; love story, the characteristic should be affectionate; and for a drama, the character should be serious and determined. Look for inspiration by observing life around you (friends, mall-goers, pedestrians) and don't design in a vacuum—do your research.

Use known conventions (archetypes, caricatures). Use Friends, colleagues or acquaintances as a source for developing your character(s), they are your best resources. You know them well enough, to pick the nuggets, best part of their personalities. Do your research, write down specific characteristics/traits such as clothing, speech, tattoo(s), culture, language, accent—anything to give your character depth. In the classic 1939 film *The Wizard of Oz,* Dorothy Gale of Kansas met some friends in the Land of Oz who reminded her of friends back home. Dorothy wanted to go home; the scarecrow wanted a brain; the tin man, a heart; and the cowardly lion, courage. These motivational features can be expanded to any situation. What are some motivational traits of people you know or imagine? Continue your research and observe people and guess what they do for a living and what are their motivation/dreams?

Fig 7.7

Computer generated character by Robin Catania

V. Digital Character Appearance

It is important, in the storytelling process, to understand the framework for character development. Consider the complexity of creating the a computer generated character (Figure 7.7). Authors, filmmakers, playwrights, scriptwriters, and game developers all use sketch artists to make composites of the characters' appearance, clothing, and props. The concept art is scanned or created into a digital format so it can be manipulated as the story dictates. Consider the appearance of character and visual effect of Vadar in the George Lucas film and then see the parody directed by Mel Brooks in *Space Balls.* Characters can

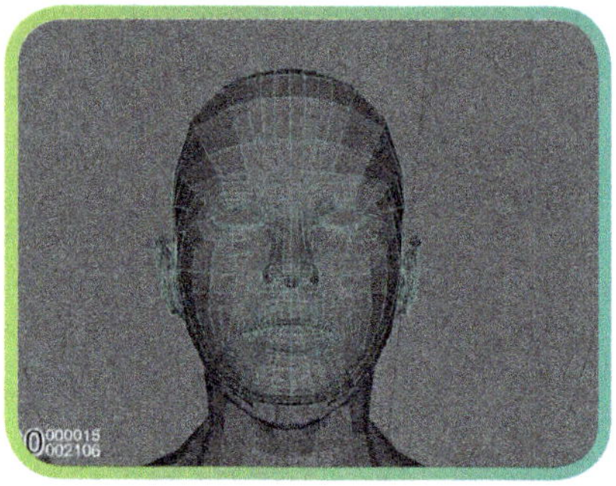

Fig 7.8

Wireframe of computer generated character by Steven Brown

be portrayed differently when a story is adapted from one medium to another. Take for example the book *Charlie and the Chocolate Factory*, written in 1964 by Roald Dahl and the film adaptations made with Gene Wilder in 1971 and Johnny Depp in 2005. The plot of the story remains intact: that of a poor but perpetually optimistic child who, along with four other children from around the world, win a golden ticket and tour of the famous Wonka chocolate factory. Examine the personality, appearance, and actions of these distinctly different characters.

Next consider different media, resources, technologies, and human effort that go into designing a digital character (Figure 7.8). You as the writer need to go into great detail to describe the characters in your story. Each time a story is adapted, the character's "mojo" changes. Case in point is the story *A Christmas Carol* by Charles Dickens. The main character, Ebenezer Scrooge, is introduced as a mean-spirited person who changes after he is visited by three ghosts. Since the original, film made in 1938, there have been six adaptations. In the original, starring Reginald Owen, the story was shot in black and white, while the latest version produced in 2009 staring Jim Carrey is a 3D animation.

VI. Point of View

Choose the point of view of the digital story from the first, second, or third person. It is the lens that the author uses to unfold the story. Non-linear stories such as autobiographies and memoirs use first-person POV. Many writers choose the third-person POV. The POV is the way the author describes what is going on in the story. According to dictionary.com, point of view:

Noun

1. A specified or stated manner of consideration or appraisal; standpoint; *from the point of view of a doctor.*
2. An opinion, attitude or judgment; *He refused to change his point of view in the matter.*
3. The position of the narrator in relation to the story, as indicated by the narrator's outlook from which the events are depicted and the attitude towards the characters.

The best practice for creating the story's POV is for the author to immerse him or her so deeply in the character's personality to embody the thoughts, emotions, and intelligence and become that person. Delve into the mind of the character. Many writers believe that developing a POV is an organic process and is best created by just writing, so the story is truly authentic if it comes from the heart of the author.

In interactive digital stories, the end user can have a role as first person or third person, depending on the type of game. A player sees through the eyes of the avatar (a representation of the player) in a first-person POV. Some game developers use cut scenes and pre-scripted sequences to shift perspective so that the avatar can be seen in third-person POV.

A. FIRST-PERSON POV

The first-person POV in books, film, TV, and video presents the story through the main character's eyes or voice. The story can have more than one person telling story. This POV uses the pronoun "I" as singular and "we" in the plural. In the last scene of the film *When Harry Met Sally the* narrator moves from the singular "I" to the plural "we" as the older version of the couple recounts the way they first met. First-person POV allows the reader/viewer/game player to see the point of view from the character's perspective, hear their opinions and thoughts, and feel their emotions. Some first-person digital narratives rely on information that is given or retrieved through external sources such as an overheard conversation. Authors such as Pat Conroy in *Prince of Tides* and Jack Kerouac in *On the Road*, use the first-person POV in their novels. Films in the first person include the *Blair Witch Project* and the *Lady in the Lake*. In gameplay, the player sees and reacts in the first person, similar to real life. Games of first-person shooters include *Mirror's Edge* and *Skyrim*. The winter holiday classic, 1983's *A Christmas Story*, by Jean Shepard, is told in the first-person POV. in the book *The Big Fish*, written by Daniel Wallace, the story unfolds in the first-person point of view by William, Edward Bloom's son. In the film adaptation, Edward Bloom tells the story. Radio talk show host Garrison Keillor takes on the persona of Guy Noir, private eye, in his weekly talk radio show on National Public Radio (NPR). A few examples of books written in first-person POV are *Jane Eyre* by Charlotte Bronte as well as *Black Beauty* by Anna Sewall, the *Odd Thomas* series by Dean Koontz, and *The Murder of Roger Ackroyd* by Agatha Christie.

B. SECOND-PERSON POV

This POV talks directly to the reader or viewer. The second POV uses the pronouns "you" and "yours." It is not used very often in narrative format. It is used frequently in memos and personal correspondence. In the television sitcom *Modern Family,* a member of the family faces the camera and speaks directly to the audience.

C. THIRD-PERSON POV

Third-person POV uses the pronouns, "he," "she," or "it" to inform the end user that the story's perspective is outside the characters' voices, whereas the first-person POV comes from within. Writers commonly use this POV rather than the first-person POV when referring to a person, place, or idea. Consider how many stories use this POV, from nursery rhymes such as, "There was an old lady who lived in a shoe" to Charles Dickens' *Tale of Two Cities* ("It was the best of times, it was the worst of times ..."). This approach to storytelling is very popular. Think of the perspective to the story like sitting in the same room with the author as he/she explains the story to you. The end user is an outsider, an observer. The third-person POV differs from the first-person POV because the author's voice not the character's voice is what is heard. In literature many books have been written in the third-person perspective including *1984,* a science fiction narrative by George Orwell.

The Great Gatsby is told in the first- and third-person POV. This story by F. Scott Fitzgerald is told through the narration of Nick Carraway as he alternates between first and third person POV as the scenes appear to him throughout sequences of the story. This story has the Carraway character romanticizing and admiring while also admonishing Gatsby. The events of the story are described in nostalgic and melancholic terms. Third-person

POVs can be developed subjectively that rely on descriptions of a particular character's thoughts and emotions as Carraway did through his journey to get to know the Great Gatsby. The objective approach gives the author agency to move from character to character instead of remaining on a single character. This approach also gives the reader/viewer knowledge, which is often told in an omniscient voice. An omniscient approach, according to Wikipedia.com, states, "An omniscient narrator has knowledge of all times, people, places, and events, including all characters' thoughts; a limited narrator, in contrast, may know absolutely everything about a single character and every piece of knowledge in that character's mind, but the narrator's knowledge is 'limited' to that character—that is, the narrator cannot describe things unknown to the character."

H. WRITING DIALOGUE

The dialogue must fit the media. For example, ask yourself what the differences in dialogue are between print, film/video, performance, and interactivity. Then ask yourself what the similarities are. In deciding what dialogue style to choose from, consider the ways effective dialogue can differentiate major characters from one another. Gradually, as they talk, you should hear differences begin to emerge. Consider the character's choice of words and the way they talk. Is there a colloquial quality to their dialect? The choice of words can really establish a character's identity. Key words or phrases, diction, accent, slang, code phrases, sentence structure, profanity, even the amount of dialogue—all of these can tell the audience a great deal about the persona of the character(s) you are creating. In order to develop these voices, practice writing and speaking fictional conversations between characters in the narrative and game. Explore the way the characters talk in a hypothetical conversation that takes place in their world. A way to create effective dialogue is to act out the voice of the character and literally talk their talk. In game characters, the conversation isn't actually going to be featured in the gameplay. It's an asset that the author is creating as part of the game design, but it's not intended for inclusion in the game's dialogue. Also, you should write as though the two characters are talking to each other about a topic of significance in your game world. Let the characters converse as you write, first in one character's voice, then in the other's. Explore the characters' present circumstances, including

- Their cultural background
- Geographic background—The South vs. New England—accents, mannerisms, things that are proper to say
- Age—intelligence, experience
- Flaws or experience—sexual abuse, fat, divorced, etc.
- Their goals, intentions, and dreams—what did they get versus what didn't they get
- Sobriety—of sound mind
- The topic being discussed—triggering something in their past

In developing dialogue for interactivity create a dialogue tree similar to flow chart in (Figure 7.9). This is useful to distinguish choices that

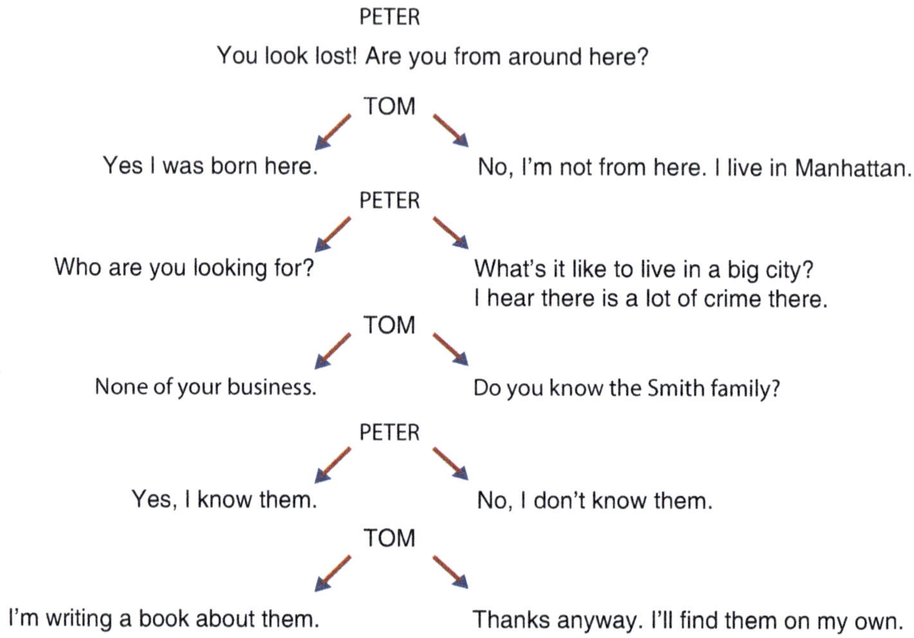

Fig 7.9

Dialogue tree

the characters are faced with and what their responses to each choice would be. Dialogue trees appeared as part of the story structure as early as 1941. *Run Lola Run* is an early example of Interactive film. This 1998 German crime thriller is the story is about Lola, who needs to obtain 100,000 German marks (50,984 Euro) in 20 minutes to save her boyfriend's life.

For examples of character dialogue for games, review Gamasutra.com. This provides helpful tips to developing game-driven dialogue. The method of writing dialogue is different from print, film/video, and interactive media. The format of the dialogue is different between these mediums. In writing dialogue for print, the conversation is part of the narrative. It gives the characters personality makes them come alive.

In writing for film, television, and interactive narrative, the dialogue structure is different. The story becomes more interesting because the characters are revealing their personalities and idiosyncrasies. An effective method is to select characters that have strong personalities, opinions, and philosophies. This is a way to introduce conflict or disagreement of some kind into the narrative. In adapting a story from print to film, there is less time for exposition, so the effort needs to go into the telling of the story. Show, don't tell, and make sure the dialogue fits the story.

- Focus on what is happening
  - Sound
  - Sight
  - (Sometimes) Smell, Feel, Taste
- Write what you think your character should say
- Talk as if it was you and what you see (and experience)
- Work on their voice
- "Hear" the words in their head
- Don't write proper sentences
- Make sure they speak "in character"
- No big words from toddlers, scientists may drop science-y words, etc.

Read out loud what you have written and determine how the dialogue flows from the character(s). Ask yourself:

- Does it sound right?
- Is it natural?
- Is it too long?
- Make sure the exchanges are brisk
- Is it just the right amount of spoken words or is it too short or long?
- Are you using the simplest words possible to get the message across?
- No "monologues" unless you are writing as Shakespeare would

I. WRITING THE SCRIPT

a. Master Scene Scripts

Scriptwriting is the blueprint of the scene in the story with information about dialogue, location, camera moves, and angles and lighting. It is essential to any narrative. Documentaries and non-fiction stories are unscripted. The story unravels through interviews. Voice-overs are usually part of documentaries. A voiceover is a narration that accompanies the visuals in a film or video. The narrator is usually off camera in a documentary, with the exception of documentary filmmaker Michael Moore. He is usually in his films and also does the voice-over (VO) in *Bowling for Columbine* (2002) and *Fahrenheit 9/11.* Scripts provide the cast and crew with:

- Info on characters, locations, stage direction, and dialogue
- Helps the director clarify ideas and develop a project that effectively delivers a specific message to the target audience
- Helps coordinate the production team
- Helps formulate the budget

Sample scripts and free scripts offered by Project Gutenberg.

A script will involve just four elements:

1. **Sluglines:** also known as Scene Headings. These appear at the beginning of a new scene and tell us the scene's setting.
   INT./EXT. (Interior or exterior) Camera location
   STREET/LIVING ROOM Action location—Short description
   DAY/NIGHT (Time of Day)—(occasionally DAWN/SUNSET)
2. **Action:** This describes what is happening on the screen and which characters (if any) are involved. Action follows standard rules of capitalization. It's single-spaced and always in present tense. (If the action happened in the past, the slugline will tell us this.)
3. **Character Names:** This always appears above dialogue and tells us which character is speaking. Character names are always in ALL CAPS.
   And sometimes you'll have minor characters that you won't want to name. It's okay to just call them SALESCLERK or PEDESTRIAN or MONKEY MAN. If there are several of the same types of character, add a number: COP #1 or BODY BUILDER #2.
4. **Dialogue:** The words the character speaks. Dialogue is centered on the page under the character's name, which is always in all capital letters when used as a dialogue heading. If you describe the way a character looks or speaks before the dialogue begins or as it begins, this is typed below the character's name in parentheses.

Audio is script:

- Two types of sound:
  - Diegetic—the characters in the story **can** hear it—used to build character, give a sense of time and place
    - i.e. Music on the radio in the scene
  - Non-diegetic—the characters in the story **can't** hear it, but the audience can—used to build tension or set a mood
    - i.e. Soundtrack music, in most cases.

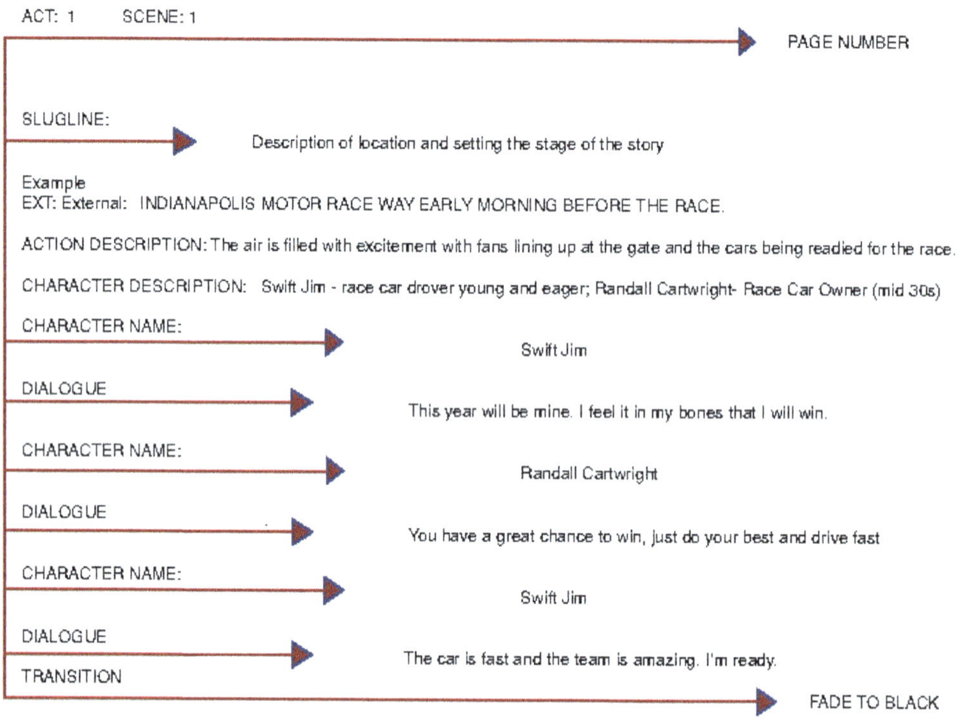

Fig 7.10

Example of a master script scene

For film, television, and video scripts, also include audio, camera angles/ moves, and lighting. For stage production, a script would include props and stage directions for scene changes. The script for a shot, scene, or sequence for film, video, and stage follows this standard format:

Master Scene—Standard format

There are many commercial scriptwriting software applications. Here are a few:

- Scriptwriting:
  - Final Draft
  - Movie Magic—Pro screenwriting software for Mac and PC

Adobe Story—Online (parts may be free)

BPC-Screenplay—Screenwriting software for Windows

Celtx—Open source for Mac, Windows, and Linux. Easy to use and great formatting options.

Montage software—Popular screenwriting software for Mac only.

Page 2 Stage—Open source screenwriting software for Windows.

- Story Outline and Development:

  Dramatica Pro

  Contour

  Save the Cat

- Both:

  Movie Outline—Screenwriter and story development software (Mac and Windows)

  Montage—Just for Mac OS X

  FiveSprockets-vScripter—Web-based story-development and script-writing software.

TITLE: **SCENE:**	**DATE:**
VIDEO	**AUDIO**
Abbreviated logo animation plays over various studio audience shots [WS]—wide shot [MS]—medium shot [CU]—close-up shot [WS] sweeps the studio featuring logo, showing TALENT Post super: *[talent names]*	*[CUE MUSIC]* *[MUSIC AND APPLAUSE CONTINUE]* TALENT: Welcome to Digital game show! Let's meet the players and get started! **[1:15]**
WS of talent at game base Cut to MS OF contestant #1 *GAMEPLAY: SEMI-FINAL, GAME ONE, FIRST TURN*	TALENT: Everybody wins; however, the goal for these players is to win enough to advance to the next round. **[AUDIENCE APPLAUSE] [2:30]**

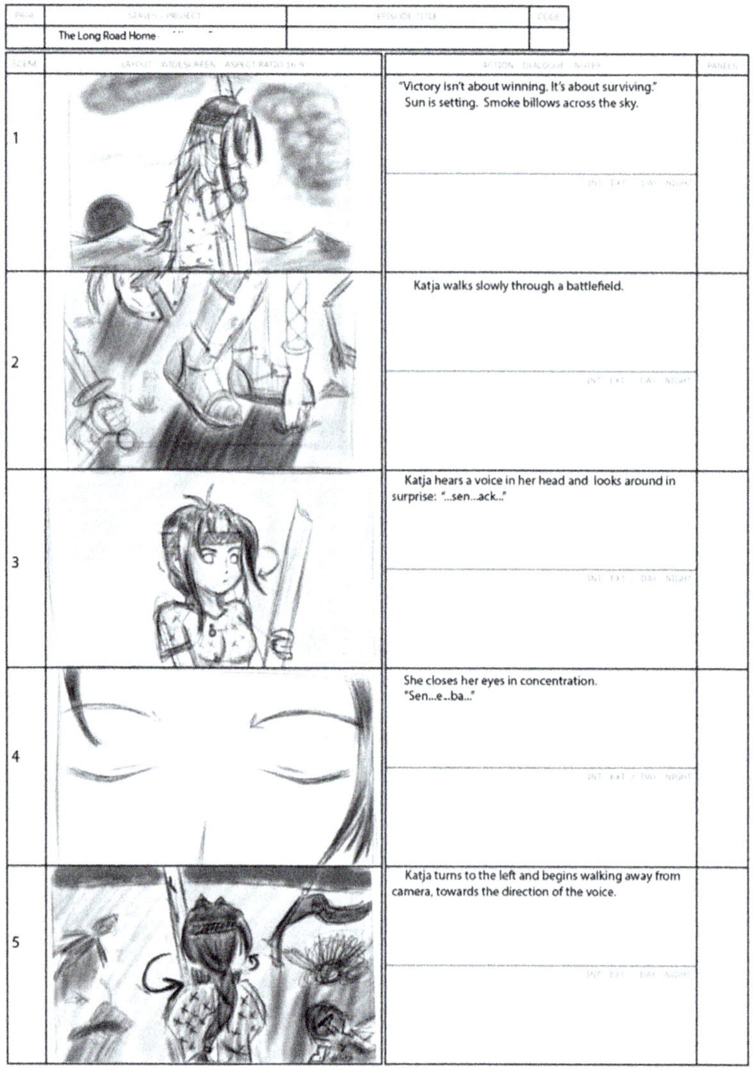

Fig 7.11

Storyboard by Elizabeth Watness

b. Dual-Column Scripts

Creating a TWO-COLUMN script differs from the master script in that there is a column for video with a corresponding column for audio. A sample two-column script is less detailed than a master script.

J. CREATING THE STORYBOARD

The purpose of creating a storyboard is to map out the story into sequences/scenes or chapters (Figure 7.12). It is an organizational tool that is used to separate the story into a sequential format with visuals and text. This is used as a method to see the plot of the story before beginning production. This includes film, animation, motion graphic, video, and interactive media. Most commonly, storyboards are drawn in pen or pencil. If you don't like to draw, you can also take photos, cut out pictures from magazines, or use a computer to make your storyboards. Keep in mind that your drawings don't have to be fancy! In fact, you want to spend just a few minutes drawing each frame. Use basic shapes, stick figures, and simple backgrounds. If you draw your storyboard frames on index cards, you can rearrange them to move parts of the story around.

Walt Disney used storyboards in his classic animations. He used storyboards to see how each character acted over a sequence of actions. He used storyboards to refine the story before progressing to the production stage. Storyboard artists were employed to draw the parts of the story that are key, where the action changes. Storyboards, digital or traditional (hand drawn), are formatted with a sequence of visuals on top of text that explains what is happening in the frame. Classically, storyboards were stacks of thumbnails (small drawings) pinned to a wall and accompanied by a presentation; they can also be a document of thumbnails, or even something that looks more like a comic. Once a script is written for a film or animation, the next step is to make a storyboard. A storyboard visually tells the story of an animation panel by panel, kind of like a comic book.

To make a storyboard, use the white squares for the key frame images from your story. Begin by deconstructing your story and using your

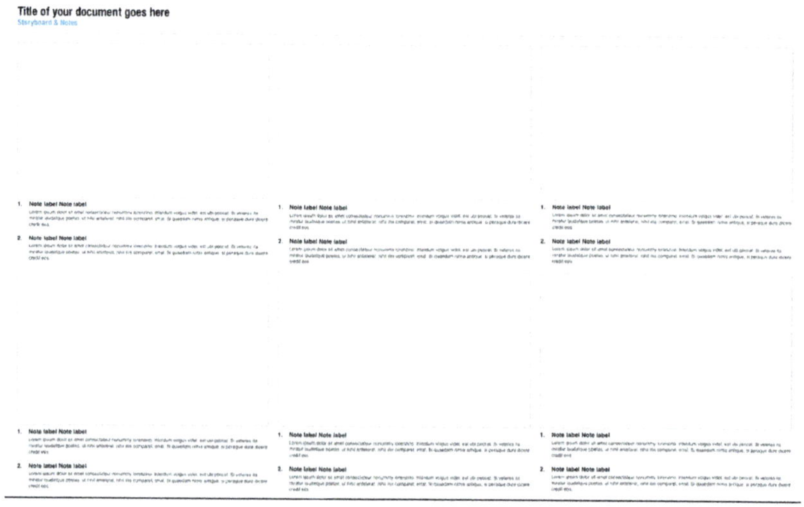

Fig 7.12

Storyboard template

outline to create a storyboard that shows **the highlights, key frames** of your story. The key frames are the main events in the story. Use the lines under the white squares to make **notes about your story**. These are brief explanations of what is going on in the drawing. In film, video, games, and other time-based stories, there are arrows indicating camera shots (Figure 7.12). These are some of the most common abbreviations of camera shots:

ECU or XCU—extreme close-up
CL—close-up
MS—medium shot
MLS—medium long shot
LS—long shot
ELS—extreme long shot
FS—full shot
WS—wide shot
OS—over the shoulder

Oftentimes a storyboard artist will include *motion arrows* (Figure 7.13). These arrows are intended to show how the camera is moving in that shot. It's a powerful way to indicate any extra action that is taking place

Fig 7.13

Storyboard with camera directions

in the frame. Storyboards usually are created after the script has been written. Storyboards are used to:

- Define story flow
- Create key frames that show how the action in the scene changes
- Refine the story before production
- Allow the authors, filmmakers, game developers to see the action of the story and also what the characters will look like prior to production. Storyboards are created in pre-production.

Storyboards represent what the camera sees rather than the "idea," which is created in the concept-development process. Each frame of the storyboard has appropriate notes for each shot, detailing what is happening and how the shot is taking place. They are usually rough sketches that focus on *rough rendering*.

The idea is to just draw it well enough to convey the idea of the shot. In other words, stick figures, photographs, or drawings are perfectly adequate for storyboards (Figure 7.13).

Once finished with a storyboard, take a step back and look at it and ask yourself:

- Is it conveying the entire story?
- Could someone take this document and recreate the shots accurately?
- Are there missing shots that are needed to explain going from point A to B?
- Are there any rough spots—areas that are confusing or take up too much time?

Your storyboard should convey some of the following information:

- Which characters are in the frame, what are they doing?
- What are the characters saying to each other, if anything?
- How much time has passed between the last frame of the storyboard and the current one?
- Where is the "camera" in the scene? (Close or far away? Is the camera moving?)

The flow of the story should be mimicked in the storyboard. Looking through a storyboard should give the viewer the general idea of what is happening within the shot and the story quickly to realize that part of the story doesn't fit or isn't working right; this can oftentimes be caught in this stage, and reworked. The animator, writer, and film producer Nick Parks, creator of the *Wallace and Gromit* stories, is known to act out the frames of the storyboards for his films. Creating a storyboard will help you plan your digital story frame by frame. You can make changes to your storyboard before you start production, instead of changing your mind later. You will also be able to show your storyboard to other people to get feedback on your ideas.

Other uses of storyboards are:

- Business process
    - Storyboard to show what people will do, when they will do it, etc.
- Novel writing
    - Helping establish shots and mood
- Comic book writing
    - Comic book writers will often draw smaller versions of what they are working on
- Interactive media design
    - Detailed steps users must go through in order to accomplish some task
    - Shows the user interface

Below is a list of software that sets up the script according the appropriate medium's storyboard templates:

Storyboard Quick—video
StoryboardPro
Frame Forge (3D)
Springboard
Storyboard That
ScriptVOX Studio
Director's Notebook
Moviestorm
Storyboard Tools

The best method to learn the importance of creating storyboards is to look at the work of excellent storytellers. The rule is we learn from our past, so my advice to you is to revisit some powerful storytellers.

K. ANIMATICS

Animatics can easily be described as animated storyboards. They are 2D and 3D hand drawings that are scanned into the computer or created digitally. Animatics are used to preview scenes of an ad, film, or short story before they are shot or animated, and before money is allocated too heavily on a shot that may not make final production. Animatics are produced by filming physical storyboard drawings. Animatic production has since become digital and much more streamlined. Animatics often follow the storyboarding practice, but sometimes storyboarding is skipped when the project already has the final audio dialog and scratch tracks. A heavy sense for the project's timing and overall success is the primary goal while in this stage of preproduction. Entire budgets for films and shots have been won and lost on the ability of the animatic to portray the director's vision both simply and effectively.

An animatic is simply a series of still images edited together and displayed in sequence with a rough dialogue and/or rough soundtrack added to the sequence of still images (usually taken from a storyboard) to test if the sound and images are working effectively together. As illustrated in (figure 7.14) an animatic can help plan out action sequences in a film, video or game before production. They are also used heavily to

Fig 7.14

Animatic by pengWu

assist visual effects artists and supervisors when communicating the needs of particular high-budget sequences.

Animatics assist the director and department supervisors to communicate their vision of the story in terms of timing, continuity and pacing. The storyboard and soundtrack are amended if necessary, and a new animatic may be created and reviewed with the director until the story is perfected. Editing the film at the animatic stage can avoid animation of scenes that would be edited out of the film. Animation is usually an expensive process, so there should be a minimum of deleted scenes if the film is to be completed within budget.

The storyboards are often animated to simulate real-life camera moves, such as zoom and pan. These animations can be combined with reference video, other animatics, sound effects, and dialog to create a presentation of how a film could be shot and cut together.

I. Digital Effects

Applying digital effects is the process of using computer-generated imagery (CGI) and high definition sound to give the digital story visual and contextual depth. Among compositing software, I recommend for an independent videographer Adobe After Effects for video and Audition or Pro Tools for audio. Compositing is similar to an orchestration in that each instrument (medium) adds its unique characteristic to the symphony (story). If special effects, 2D, 3D animation, motion graphics, audio enhancement, or other digital effects are used, it must not be because they add the "wow" factor. If this is the motivation, then the story is in trouble. The story needs a solid plot with engaging characters and dialogue. Adding digital media to the story should be used to enhance what is already a solid plot. Digital compositing should be used appropriately; that is, to add value. The digital story has to examine the parts of the composition and determine how each element adds to the story. The digital composition is full of choices that can make a story jump off the screen and appear as if the characters truly interact; *Forrest Gump* is a composite of live acting with existing footage.

Fig 7.15

Battleship of the Potemkin poster
Source: http://commons.wikimedia.org/wiki/File:Vintage_Potemkin.jpg. Copyright in the Public Domain.

II. Montage

Montage is an example of a composition framework that switches the scene from one to another in order to heighten the story. A classic example is the 1925 film The Battleship of the Potemkin (Figure 7.15). In one scene, The director, Eisenstein cuts between scenes of a distraught mother watching her child cascade down a flight of stairs in a baby carriage as the Russian soldiers march up them. A similar scene is in the 1987 Brian De Palma digital film The Untouchables, where Eliot Ness, played by Kevin Costner, saves a baby from similar circumstances.

III. Digital Composition

The compositing of digital media can involve Chroma-key or green/bluescreen technique. The process involves an actor performing before a green or blue background and the background is removed in post-production using digital tools such as Adobe After Effects. The green screen is replaced with another background. Films such as *Castaway, The Perfect Storm* and Bloody Omaha use this technique. The reason for the green- or bluescreen is that neither color is in the human skin pigment.

Fig 7.16

Green screen composition

Academy award for special effects was given to Arthur Widmer for his work on an ultraviolet traveling matte process. He also contributed to the development of green/bluescreen techniques (Figure 7.16). One of the first films to use this effect was the 1958 adaptation of the Ernest Hemingway novella, *The Old Man and the Sea*, starring Spencer Tracy. The 1958 *Attach of the 50 Foot Woman, King Kong* (1933), *405 the movie, Bloody Omaha, Jurassic Park* (1993*), Prestige, Matrix* and *The Empire Strikes Back* are all examples of green/bluescreen composition.

L. PRODUCING THE DIGITAL STORY

If you have completed the steps of the digital story structure, then and only then are you ready to produce your story. I encourage you to get feedback from as many people as are willing to read your work. Whether in an academic environment or competition, everyone needs feedback, constructive criticism. Every writer can always improve. Review and check off what you have done to this point in the digital story structure:

a. Planning and Organization
b. Writing the Concept Document
c. Writing the Treatment
d. Creating the Outline
e. Writing the Digital Story
f. Digital Story Structure
g. Character Development for Digital Storytelling
h. Writing Dialogue
i. Writing the Script
j. Creating the Storyboard
k. Animatics
l. Producing the Digital Story
m. Dissemination of your digital story
n. Celebrating with applause

Your story has gone through pre-production, all the steps leading up to the realization of your story in digital format. Before you begin the production of taking the story further, it is a good idea to create a production schedule and become familiar with any new software you may need to produce your digital story. There are many considerations in producing a digital story. Let me focus on text-based and audio/video digital stories.

I. Formatting the Story for Electronic Devices

Text-Based Production

The digital book format has not changed very much from the traditional since the first eBook. The layout is basically the same: Table of Contents, Introduction, Body of Work, Index, and References, if an electronic textbook. The text should be formatted as a .PDF or website with security so your story is copyright protected. If your digital story was created in Microsoft Word or any word processing software, convert it to a .PDF file in Mac by going to the print option and selecting the PDF option or save it directly as a Web page under the file Save As Web Option. Production of text-based stories requires the author to format the text and images. Adobe InDesign is an application that can be useful in the page/screen layout of visuals and typography. If you plan to publish your story online, consider resizing the layout for different screen sizes, from smart phones to desktop monitors. In addition, consider creating an audio version of your digital story. Text-based books such as Design Like Apple, the a biography by John Edson about Steve Jobs and his vision for the empire he built, is available in hard copy and online versions. It is common practice for books, narrative or non-fiction, to give end users a choice.

Audio/Video/Production

If you are producing a digital story for video or film, use your script and storyboard as your blueprint. The first step is to create the media by shooting the visuals and recording the audio. There are many choices in the multimedia creation process. One is to synchronize audio and film/video recording of actors'/characters' voices in real time. The other option is automatic dialog replacement (ADR) used often in games, film and video. It is a lip-sync process often used in animated films such as *Aladdin* with voice overs by Robin Williams in which the actor has his or her voice recorded in synchronization with their animated character on the screen. Audio for film and video include background sound. If you plan to create your own soundtrack, make it the same quality as the video. If you are producing a high-definition video (HDV) story, the frame size should be 1920h 1080v; frame rate of 24 or 30 frames per second (fps); 60 frames per second if you are shooting sports or want to capture a slow-motion scene; and sound sample rate 48 kilohertz (kHz). The pixel aspect ratio can either be square or 4:3 ratio or HD 16:9. This is the size of the screen the story will be displayed on. IIt is best to produce your story using the highest quality resolution, which currently is 1920 x 1080. Keep in mind Moore's law about technology that it will change every 18 months. Royalty-free software is available online as a choice if you are unable to produce your own background soundtrack. If your story is non-narrative, such as a documentary, it is common practice to use voiceover (VO) that is recorded after the visuals are created.

II. Equipment

A digital story can be created using a smart phone equipped with recording audio, visual, and text media. If you want to use a more sophisticated camera, consider the DSLR, high-end photographic cameras with HD 1080 video capability. There are many choices for which camera to purchase. I would look at one that will last you 3–5 years and of course be in your price range.

Using a Digital Camera to Create Visuals and Audio

Most digital cameras and smart phones are easy to use to record a story. Some suggestions are to frame the shot so there are no distractions in the background of the composition. Check the lighting conditions. How would you work with exterior scenes if the day is very bright or during night filming when there might not be enough light? If you are

Fig 7.17

Three point lighting

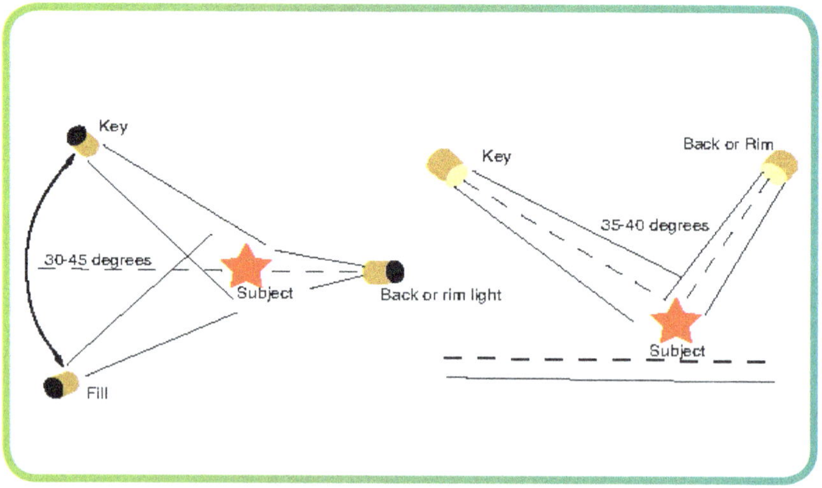

in a studio set-up, use the three-point lighting set-up that illustrates the position of each light. The key light is the main light and is used to enhance the subject. The fill and back or rim light add dimension and create a separation of the subject from the background (Figure 7.17). The key and fill light should normally be placed approximately 30–45 degree angle. from the subject The back or rim light should be placed behind the subject at steeper angle than the key and fill lights. If you want to make your story more professional, then you need to invest some money in equipment and become familiar with what is used in the industry. This will involve a cost to you if you are investing in a camera other than your smart phone. Currently the market has an abundance of cameras from point and shoot to those with interchangeable lenses such as DSLR (digital single lens reflex). Your digital story should take into account the scene composition and how the author/screenwriter uses it in the pacing, timing, and unraveling of the story. Some non-narrative stories such as documentaries and biographical stories utilize the primary footage, (A-roll) which incorporates interviews and intertwines these with B-roll, or supplemental footage. This process gives the story a more interesting approach because it avoids the "talking head" syndrome. This is when the person(s) of A-roll talk throughout the story. Ken Burns,

Fig 7.18

Screen shot of editing production application

the filmmaker and great storyteller uses the A-roll and B-roll story structure in his documentaries.

III. Understanding Time-Based Media and Software

There are a number of editing program applications such as Adobe Premier, Flash, or Apple Final Cut Pro Most editing software window have similar functionality but are arranged differently (Figure 7.18).
It is important to set up a workflow and production process to organize your media and make best use of your time. Editing takes time like all the productions steps and should be considered in the timeframe of the production.

The final format of your story needs to be considered in the concept development stage. If your story is a video consideration will occur in the export choices. Currently these include quick-time movie (.MOV); H.264 or MPEG 2 for Blue Ray and DVD. If you are considering creating an audio story, a great source for inspiration is National Public Radio (NPR) *A Prairie Home Companion* by Garrison Keillor. If your story is translated from one language (English) to another, such as Greek be very accurate to match the spoken word with the text on screen. An example is Food of the past, nourishment for the future. A short documentary, The Aesthetic Moment documents the sculpture on the campus of IUPUI, Indiana University–Purdue University Indianapolis, Indiana.

Organization is key and creating a production schedule and shot list is essential. All things at this point need to be considered in order to move your digital story from concept to dissemination. Consider how you organize the creation of the story including drawing up a daily production report.

PROJECT:		DAY:	DATE:
PRODUCTION OFFICE:		FIRST SHOT AM:	CREW CALL:
		Meal 1 In:	Meal 1 Out:
P:	P:	Meal 2/Break In:	Meal 2/Break Out:
F:	F:		
			Wrap:
Locations:	Production Crew:		
1)	Executive Producer:		
2)	Producer:		
3)	Director:		
4)	Production Manager		
5)	Production Coordinator:		
6)	1st Assistant Director:		

Cards Used	
Previously Shot	
Shot Today	
Total to Date	

CARD NUMBER	DESCRIPTION	AUDIO NOTES

EQUIPMENT/TECH NOTES:

CREW/TALENT:

POSITION	NAME	TRAVEL	TIME IN	TIME OUT	OVERTIME	EXPLANATION FOR OT
Director						
Producer						
PM						
PC						
1st AD						
2nd AD						
DOP						
Cam Assist						
Sound						
Gaffer						
Key Grip						
Makeup						
Wardrobe						
Set Design						
Craft						
PA						
PA						
Talent						
Talent						
Talent						
Talent						

M. DISSEMINATING THE DIGITAL STORY

I highly recommend that prior to publishing your digital story, no matter the type, have someone read, listen to, or watch the work one last time. Do not release it until you are absolutely certain of it, because once it goes online, it is out there. It might even go viral if you are lucky. There are many online self-publishing websites for posting a text-based story you have written. Make sure the one you choose is reputable and has a publication history. If you plan to make money from your story, it is best to have a contract with the publisher. If you are disseminating an audio story, consider creating podcasting. Keep aware of any copyrights that you might need for protection. Digital stories often have warnings of copyright protection and watermarks to assist in the security of your digital story. For video, free sites include YouTube, Vimeo, and social media, Facebook, etc. Consider competitions and calls for submissions in festivals. Look at publications, calls for entries in the context of your digital story and enter those. Try local or regional before you submitting to a national or international venue. It is important to try again even if you don't get accepted at first. Continue to write because you will improve with each digital story you write, and most importantly, don't be afraid to think outside the box.

N. CELEBRATING WITH APPLAUSE

You are done with your first digital story. It is indeed time to celebrate your accomplishment. Be patient with the results. Comments and constructive criticism are important feedback for you to use as you develop and improve your skills as a digital storyteller.

EXERCISES FOR STUDENTS

a) Research different types of stories with examples.
b) Research stories with similar themes and discuss.
c) Discuss the Development Process.
d) Work individually or in a group and complete all the steps of the Development Process.
e) Discuss the Hero's Journey.
f) Research competitions for you to submit your digital story.

3 Part Three
INTERACTIVE STORIES

8. OVERVIEW

A. DEFINITION AND TERMS

Interactivity is a signature feature of digital storytelling. Games are the prominent form of interactive stories. They date back to ancient times such as Greek mythology in which stories arose about Zeus (Figure 8.1) controlling mortals. Interactivity is a tool for building healthy competition, skills, and critical thinking. Most interactive stories have a clearly defined goal, follow specific rules, and have challenges and rewards. Competition builds excitement and interest in these types digital stories.

Fig 8.1

Zeus

Source: http://en.wikipedia.org/wiki/File:Jupiter_Smyrna_Louvre_Ma13.jpg. Copyright in the Public Domain.

I. Agency and Pacing

Interactivity allows or relates to continuous two-way transfer of information between a user and the central point of a communication system, such as a computer, game console, mobile device, or television. More simply put, the viewer in a traditional story becomes a player, gamer, end user in an interactive one. The user is given agency or the ability to take actions within the digital narrative. A game allows the user limited choices such as driving a car instead of walking, climbing a tower, talking, and trading with people. In giving the end user agency, be aware of the pitfalls. "With great power comes great responsibility," wrote Stan Lee in his comic book, *Spider-Man*. In terms of developing an interactive story, giving the user more agency requires more planning, and with every choice comes the need for more layers

Portions of this section of the text were contributed by Travis Faas.

of content. Giving players too much agency, the ability to do a lot of things, allows them to do things they hadn't anticipated. These unintended events can be delightful or aggravating. Giving the user the ability to do whatever they want can potentially lead to a fragmented story.

As a story becomes more specific, players must be offered less choice. Giving a player fewer agencies restricts their freedom but also allows the storyteller to be able to force a player into choosing the "right" choice to continue. Great examples of games where players have little agency can be found in the "point and click adventure game" genre. These are games with very linear plotlines where typically only one choice is the correct one to move forward in the game/plot. The job of the player is to think through their options, trying to figure out the right choice to advance the story. Cheating is unfortunately example of misuse of agency. Take *Griefers;* in games, these are players who take joy in simply annoying the other players.

Not all games restrict the player's choices to be only one right option. Good examples of games on the opposite end of the agency spectrum are sandbox-style games, where players are given many options of things to do, while not forcing the player to explicitly do those tasks in a particular order. The poster child of sandbox-style games, *Grand Theft Auto*, exemplifies this abundance of agency. At any given point, the player can choose to compete in races, steal cars, look for hidden objects, create mayhem, advance the story, or any number of other things. At no point does the game force the player to do any individual thing, and many players will happily speak of the hours they spend in the game just driving around the city and listening to the radio.

Some sandbox games do not even have what would be called a story, though others will let the players take the story in at their own pace, through careful placement of missions inside the game world. When crafting a sandbox-style game, the work of the story will go into defining the world and the events that can take place within that world. The eventual goal of this type of storytelling is to enable the player to craft a personal story within the game. A well-constructed world can still impart a particular "flavor" to the player's story. For example, the story that a player will experience in *Grand Theft Auto* while stealing cars and doing other nefarious activities will be quite different than the story experienced while playing *Skyrim*, a world filled with orcs, magic, and mysterious caves. That is not to say that stories cannot be told within the sandbox context, just that they will tend to be shorter and less involved. There is no shortage of villagers in Skyrim who need something done for various scripted reasons, but the player is not required to actually help them out, and the actual dramatic content of any given quest is not likely to be very high.

Once a designer has figured out how much agency to give the player, their next task is balancing the messages coming from the story and the gameplay. When looking to craft a story into a game, the story and the gameplay should be in agreement about the current capabilities and standing of the protagonist. For instance, if the story sets up a character who is notably inexperienced in combat, it would not make sense for that character to then pick up a sword and hack away with the best of them in the gameplay segment. A good story-driven game will reinforce the protagonist's inexperience by having him be quite ineffectual during the fight, perhaps to the point of needing to run away. This works as well for narrative stories. If the story claims that the antagonist has appeared and stolen all your money, then the player should not have their money in the game when they go to look for it. In the game *Assassin's Creed,* when the protagonist is demoted from his high rank, the gameplay changes to show that he has lost his weapons and many capabilities. Linking a story to a game can be complex and for best effect requires interaction between the game designer and storyteller (should they not be the same person).

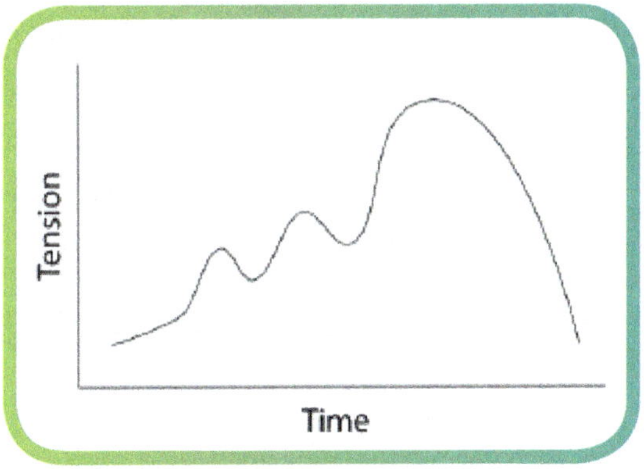

Fig 8.2

Game development

A major challenge of any game is keeping the pacing and tension (as indicated by both the story and the gameplay) in check while still letting the player control the action. For best effect, a story's tension must be equal to the difficulty in the game itself (Figure 8.2). For instance, if one is playing a cooking game and the story claims that the current recipe should be simple to make, it would be quite frustrating for the player to find out it is actually a difficult recipe. Thankfully, the common progression of game difficulty is very similar to the tension in a three-act story. Many games start out with an easy tutorial phase, which is similar to the introduction phase in a story used to introduce characters and places. After the tutorial, the game tends to get more difficult, especially as players acquire new capabilities in the game and generally get better at playing the game. The general increase in difficulty is usually broken up with an occasional break period where the game does not become harder in any way. These break periods are designed to give the player a chance to practice their skills before things get more difficult again. Following the pattern of gradual difficulty increases with the occasional break, the game will progress to a final "boss" battle designed to be the most difficult test of a player's skills. In other words, the boss battle is the climax of the game. A good game narrative would have the climax coincide with the final boss battle, having both moments of high tension reach their resolution at the end of the game.

Giving a player agency forces them to take action within the game, which has the potential to get the player more invested in the game and their decisions. A high level of investment (or immersion) can become a tool in the storyteller's belt, enabling them to tug at certain feelings based on the fact that the player chose to have a certain event happen. Showing someone being robbed on the screen is one thing; the player choosing to do the robbing can be quite different. This does not mean that everything should require a player action to make it more "impactful." If the choice is forced upon them (for example, forcing the player to press a button to commit a crime with no way out), there is a chance the player could rebel against their immersion and stop associating their game actions with their personal choices.

Games will often, either by design or by their nature, put the player in a state of mind (called "flow" or "flow state") where they are challenged but not overwhelmed. Many stories have protagonists who are not within a flow state near the middle of the narrative, when the protagonist has been defeated once and is looking for the path to victory. The difference between what the narrative is calling for (total defeat) and what the game is calling for (just another minimal challenge increase) can be hard to reconcile. Such "complete failure" plot points should not be avoided in a game's story. Instead, if the story calls for a player's total defeat, it would be best not to tie such an event to a gameplay portion. Many game players can attest to the frustration they feel when presented with a challenge that is insurmountable. While this actually can be a very powerful storytelling element (the player actively feels the frustration at their inability to win), it is a terrible gameplay element (there is little fun in a challenge that cannot be beaten), and is likely to lose players due to the frustration. Approach such moments with caution and, when they are required, attempt to mitigate the frustration by quickly introducing a new goal that is attainable for the player.

Interaction builds critical thinking skills and can become a learning experience as the characters face universal challenges and decisions that will affect the outcome of the digital story. There are always positive results when both the mind and the body are engaged in an activity; passive behavior becomes active and participatory and the mind and body are invested in the experience of building decision-making skills. Interactivity allows a narrative to go a myriad of ways:

- Points of departure and flow need to be controlled
  - Small decisions that don't affect main story points or final outcome—Can be frustrating for the user/participant
  - Keep number of key decision points low to avoid heavy branching
    - Flow is a type of immersion
- Also known as being "in the zone," it tends to happen when a series of interactions are interesting and not frustrating

These are often used in games to structure the challenge to keep players immersed. Ways of describing flow can be Text only—like a choose-your-own adventure.

"If you want to eat the taco, go to step C or page 76. If not, continue reading."

Joseph Campbell's Hero's Journey (Figure 7.3) is the basis for these interactive games:

- Final Fantasy VII
- Assassin's Creed
- Mario

The storyline in game design often are non-linear. This means that events don't unfold in tight sequential order. Often in gameplay, prominence is given over the storyline, as the end user must have a significant amount of choice, time, and methods of exploration in which he or she must meet a set of accomplishments to move the character or story forward. Tools for building interactive stories are:

a. Interface and Navigation
b. Systems for Determining Events and Assigning Variables – Logic that guides the programming
c. Assigning a Role and POV to User
d. Working with New Types of Characters and Artificial Intelligence
e. New Ways of Connecting Story Elements
f. Gameplay à fun/pleasure via challenges, rules, goals
g. Rewards and Penalties keep the user motivated
h. New Kinds of Building Blocks—not scenes, but units of possibility
i. The Use of Time and Space
j. Sensors and Special Hardware

II. Navigation

The term "hypertextual navigation" is used in interactive stories. Hypertext = extractive elements in which the user must use the computer and software to make reading choices in the database—from the entire world wide web to a particular learning package, an adventure game, or the hard drive on your PC. The result of such interactions will be that the user constructs individualized text made up from all the segments of text, which they call up through their navigation process (Figure 8.3) Listed below are types of navigations for interactive stories:

Immersive Navigation

- Representations of space = Immersive
- Immersive Media = Hypertext + Represented Space (e.g. virtual reality, simulations, and games)
- During Hypertextual Navigation we go into a site and extract information as an outsider

- However, when we move to **navigating representations of space or simulated 3D worlds**, we move into an immersive interaction, embodied by the interface

Registrational Interactivity

- To "write back into" collective information
- Examples: Updating Facebook, leaving comments on blogs, re-tweets, video responses, and wikis, etc. This interactivity depends on other users in collaboration and a system to freely allow collaboration
- Concern: User loses control of information once it is entered into a system

Interactive Communications

- Based on face-to-face human communication
- For example: A forum post would be considered less interactive than a live chat site. It's the continuation of face-to-face storytelling, basically in a digital context

Your tasks in creating interactive stories are to:

- Create a clear guidance system for the user—feedback to user, clear direction
- Design simple, easy-to-understand navigation and access routes
- Define what takes place on every screen
- Design controls for interaction—providing flexibility to the user (allow intro skip, etc.)
- Create a storyboard/flowchart to clearly show the navigational outline
- Turning information into experience:
- Create an INTUITIVE user experience that continually piques interest along the way
- Motivate interaction by providing CLEAR guidance
- Relinquish control—allow users to make navigational decisions with ease
- Keep in mind the differences between viewers/users—multiple intelligences (linguistic, mathematical, spatial, and intra-personal). Appeal to them all, if possible.

Fig 8.3

Navigation storyboard for game

III. Types of Interactivity

1. Stimulus and response is perhaps the most basic type of interactivity. The end user is shown a button and he or she clicks it, a monster appears, monster is attacked. This type of interactivity requires the computer to show the stimulus, the story and the end user to respond, and typically the computer provides some feedback: *Don't shoot the puppy*.
2. Movement and navigation open on a web page that has choices in a text adventure the player uses the WASD (keyboard in shooters) and employs avatar-based movement in which the avatar acts as the player. Movement and navigation are vital skills of living creatures, which are mimicked in games in a similar free-form manner. Massive multiplayer online games (MMOG's) use this type of interactivity.

3. Manipulation of objects interactivity is designed to act like the physical world. Players can pick up, write, draw, throw, shoot, drive, and impact any digital object in virtual space.
4. Conversation and dialogue in interactivity is like everyday human interaction, in that you ask a question and you get an answer. An early computer program that used conversation and dialogue was *Eliza*. The conversation is used to establish the vocabulary and speech patterns of the major characters in your game.
5. Sending information such as through social media in a form of interactivity. This type can range from online commentary to posting videos. It is pervasive, mobile, and ubiquitous.
6. Acquisition of items resembles real life as George Carlin describes in this video about collecting stuff. Gameplay involves players collecting swords, armor spells, badges, and knowledge.

IV. Obstacles

Keeping the interactivity away from fragmentation is similar to writing a digital story with no point. What is missing is the goal(s). If the player's mission is to get the "donut," do not give him or her free-form choices:

- Start at home:
  - Get in car
    - Drive to store
    - Drive *anywhere else*
- Stay home
  - Sleep
  - Do homework
  - Play video games

See how quickly that got fragmented ... we've lost control of the narrative. Giving users agency without telling them what to do isn't always the best way to go. By placing game-like items into your interactive story, you can "steer" the players to what you would like them to do by keeping them interested. This is accomplished by adding goals, obstacles, and rewards, and can potentially give structure to what was before an unstructured story.

V. Diagrams and Flowcharts

A diagram is a simplified drawing showing the appearance, structure, or workings of something; a schematic representation. Interactive designers create diagrams for a number of things they create, especially to visually interpret complex ones.

Fig 8.4

Interactive fowchart for game development

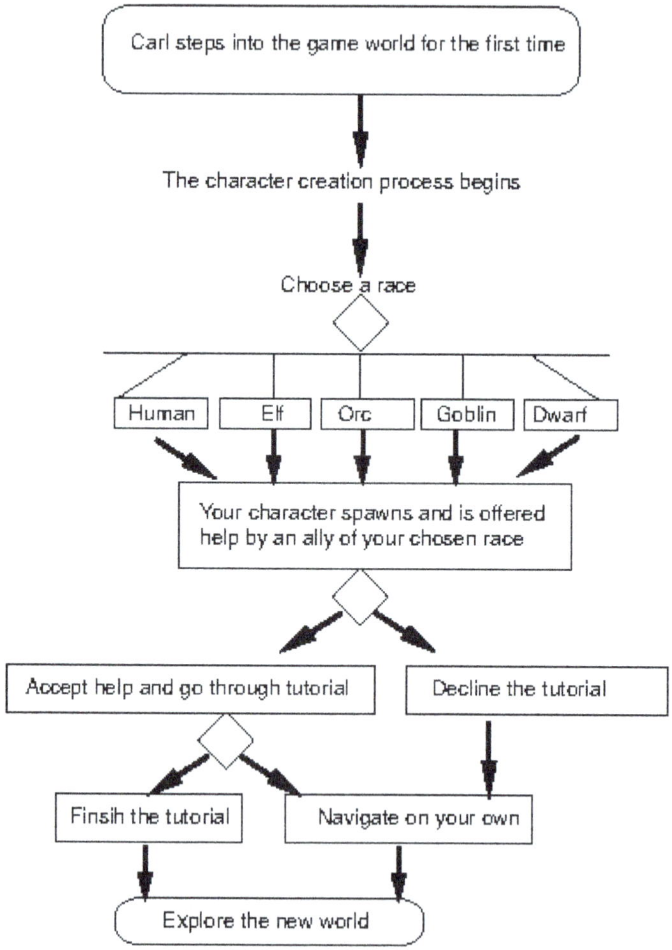

The process of diagramming helps work out kinks in the design and document the product before it goes into production.

Flowcharts are visual diagrams that display a process or strategy. Flowcharts are composed of nodes and each one represents a step in the process for the developer to analyze, design, and document the process of the interactivity (Figure 8.4).

The purpose of flowcharts is to provide an easy method to visually understand the interactivity of the story; it also plots the choices of the end user.

Flowcharts are developed with nodes that designate choices in the interactivity of the story.

Rounded rectangle: starting and stopping conditions
Rectangle: one or more actions
Diamond: decision point
Circle: connects to another flowchart
Lines/Arrows: use to connect one step to the other

- Flowchart Organization:

There is a hierarchical view that shows the entire navigation structure, a linear or one-way flow of content, and the hybrid that combines both and provides more choices of controls and layers of interactivity. Here are some examples:

- Gamasutra
- Branching
- Goldilocks
- SOFTWARE

When possible, make all your flowchart symbols the same height and width. If that's not feasible, then consider making the symbol widths the same and varying the heights for top-to-bottom flowcharts, and vice versa for left-to-right flowcharts. (Note: this doesn't apply to connector nodes and other flowchart symbols that should intentionally be smaller.)

Uneven spacing between flowchart symbols: Try to maintain even spacing both horizontally and vertically between symbols. The one exception is for Decision symbols. You should add extra spacing around these to accommodate branch labels.

Inconsistent flow direction: Overall a flowchart should have a consistent flow direction. It should be top to bottom or left to right (or right to left for RTL languages). Try to avoid mixing top-to-bottom and left-to-right flows in the same flowchart. Multi-column and S-shaped flowcharts are fine too, as long as the flowchart is consistent.

Inconsistent branch direction: It's easier to follow the logic of a flowchart if the branching directions are consistent. For example, you could make the True conditions always flow out of the bottom of Decision symbols and the false conditions flow out of the right sides of Decision symbols.

Long flow lines: If your flow lines are running from one edge of the flowchart to the other, it's better to use Connector nodes. Connector nodes are labeled circles that serve as jump points from one part of the process to another.

Too many colors: There's nothing wrong with a nice, stylistic flowchart, but don't overdo it. You don't want the flowchart's message to get lost in a sea of visual noise.

Scale: Too often flowcharts are created and then resized to fit into a single page. It's better to have them span multiple pages and be readable than to fit it on a single page and be unreadable.

Flowchart Content Mistakes

Undefined references: If you make references to other processes or subprocesses, make sure those processes are defined somewhere. Not doing so will get you into trouble if you are flowcharting for Sarbanes-Oxley, ISO 9000, or some other compliance purpose.

Poorly defined alternate paths: Sometimes processes fork. Clearly define whether there's a branch based on a decision, areas of responsibility, or whether operations can be done in parallel. Specify whether only one branch needs to be followed or all of them do.

Infinite loops: Sometimes a flowchart contains a loop-back that can lead to an infinite loop. In real practice, processes don't run infinitely (except staff meetings). There's usually some mechanism to prevent this, so make sure you document it.

Truncated descriptions: One of the biggest disadvantages of flowcharts is that they require concise process-step descriptions. However, not all operations can be described tersely. In these cases, you should use a callout, a footnote, or a separate document to go into more detail.

Flowchart Best Practices

Do not mix levels of detail: Decide whether you want your flowchart to be a high-level, mid-level, or detailed flowchart and stick with it. Anytime a section of a flowchart starts adding more detail than the rest of the flowchart, then the best thing to do is create a separate flowchart for that sub-process and link to it.

Ensure flowchart accuracy: Managers, engineers, and supervisors are usually one step removed from the hands-on experience and knowledge of their departmental functions. The people who perform the process on a regular basis should verify the flowchart steps.

Use branches instead of decisions: Decisions symbols usually represent binary (true/false) choices. Forcing a single real-life decision into a series of binary choices is often unnatural and produces bloated flowcharts—both logically and spatially.

Use symbols judiciously: Symbols have meanings, and it's good to use them when the target audience understands the meaning of each symbol. For a broad audience, though, you may be better off just using a process symbol (rectangle) for everything.

Use a flowchart key: You should consider including a flowchart key describing the symbols if you use more than a few of the basic ones (process, terminator, decision, document). What may be obvious to you may not be to someone else.

Flowchart checklist:

- Are all major elements represented?
- Are they properly and clearly labeled?
- Is there a sequence that is logical and clear to the end user?
- Are the flowchart symbols utilized correctly?

B. GAMES

Games are interconnected with digital storytelling. They inherently use similar development processes. In games the stories often use digital storytelling elements such as protagonist and antagonist character; a goal; obstacles and a series of events that culminate in the final stage of the gameplay. The process of creating a game story can be complex due to the necessity of giving the player choices and the large chance that a game player will choose to do something that wasn't meant to be part of the story. Another issue that will affect a storyteller working with games is the need to reinforce story elements with changes in the gameplay. This can become even more difficult when the game designer and the storyteller are not the same person.

Defining the word "game" can be surprisingly difficult. An imprecise assumption is that a game is "anything that is play." Before moving on to discuss the connection between stories and games, it may help to define just what "game" means for the digital storyteller. Raph Koster, author of *A Theory of Fun for Game Design*, defines games as "A form of play that has rules and a goal." Another noted games scholar, Jane McGonigal describes games as having, "... four defining traits: a goal, rules, a feedback system, and voluntary participation." The essentials of a game, taken from these definitions, seem to be goals and rules dictating how those goals can be reached (how could one begin to play checkers without knowing how the pieces moved, or when they had won?). A digital storyteller's particular definition of games might also require that the game must take place in a digital medium. For the purposes of this chapter, the word "games" can be assumed to refer to a form of play with rules and goals that takes place digitally.

I. How to Create a Game Design Document

Creating a game design document. A "concept" might be one or two pages describing the basic idea. A "treatment" might be three to fifteen pages to give a broader picture of the idea for a game. A "game design," though, is a full document (fifty to two or three hundred pages) describing a game in every detail. A game design document is a blueprint of the game. Artists, storytellers, and programmers use the game document as a development tool.

Goals and obstacles are shared building blocks of both stories and games. Not surprisingly, games tend to either generate stories of their own or have stories grafted on. There are tons of stories about heroics during a sporting match, but there also are stories of the things that go on around that match: The trials the players must overcome to make that kick, or what the win means for a family. These secondary stories that look at the greater context of the game beyond its set rules and goals are the ones that can be seen as being grafted on. The practice of grafting stories onto a game is quite common. Some great historical examples of games that had grafted stories are Pitz and Naumachia. With each example there are two stories being told: the contextual story (everything that led up to the game or what the game stands for) in addition to the story of the game itself.

The game of Pitz (which has evolved into a modern-day game called Ulama), was a brutal game that could go on for days and was played with a solid rubber ball. The goal of the game was to score more points, which were scored by hitting a wall behind the opposing team with a large rubber ball (though it may have been that the game only ended if the ball made it through a small hoop six feet off the ground that was quite hard to target). The game itself most likely could be used to tell many stories of heroic defense work and marvelous teamwork. Often the game was set within a larger social context, such as pleasing a particular god or as a replacement for actual war. While not technically part of the game itself, the grafted stories still carried great significance for the game and the players.

Another historical game that often had a story grafted atop it was the ancient Roman game of Naumachia. Because the game was staged in a flooded amphitheater outfitted with boats and captives sentenced to death, it was held fairly infrequently. In order to add an extra bit of story for the games, the boats were designed and arranged in a manner meant to recreate notable sea battles from the past or represent potential battles that could have happened at that time. The rules of the game were simple, if a bit bloody: each participant was given a team, and each team was expected to eradicate the opposing team using whatever weapons and boats were provided to them. Those who survived were typically granted pardon from their past crimes, supplying the games many different layers of story. Any of the Naumachia games could be used to tell a story of the "historical" battle, a story of the battle that took place in the amphitheater, or the individual stories of the players fighting for both their lives and freedom.

Each of these historical games had both a story they wanted to tell and a story they actually did tell. In Naumachia the players were not actually the original warriors battling on the sea, nor were their boats nearly as mobile as the originals (the boats were often tied down to one location due to size limitations). While the games may have resembled the battles they attempted to recreate, the choices of the players and the limitations placed upon them by the game itself would create a final narrative that could differ greatly from the intended one. The problem of player choice can become major frustration when it comes to using games to tell a narrative. A storyteller can take only so much control of a game before it stops being a game. By the very nature of games (being a set of rules that must be played to accomplish a goal), the players themselves must make some of the story on their own. If the storyteller removes the ability to choose, they can tell a story in just the way they want, but it won't be a game anymore.

There is a divide in opinion on just how much control the story of a game should have over how the game is played, with some game designers feeling like a game should have no (or very minimal) story elements, and others feeling as if the story elements are essential to a powerful play experience. Both are valid options. There are great games that have a bare minimum of interactivity, resembling movies with small breaks for the player to make a choice or two. It is also quite possible to make a game with a minimum of story that forces the player to create his/her own story as they journey about the world (sandbox-style games are particularly good at this). These games can still present interesting challenges for the storyteller, forcing them to focus more on world and character building and less on one particular narrative. When it comes to planning the game, most game developers have some combination of ideas, designs, and documentation that they may call the "design document." The document may take on a few different forms, and tends to have three different phases of life: the pitch, the concept, and the fluid design. Each stage may require something from a storyteller, with the final stage being the most involved of the three. Samples of design documents can be found on Tom Sloper's website.

The pitch document is like a teaser for the final game. It is very short, and only covers the high-level concept of the entire game. For this document, the storyteller may need only to define the protagonist, antagonist, key characters and briefly sketch out the setting of the story. The goal of this document is to secure funding and interest to actually produce the game. In other words, look to sell the game with its best features and don't get bogged down in small details when working at this level.

The concept document grows out from the pitch if the game secures the funding to be produced. This document will begin to go into more detail about the gameplay elements, characters, art, story, and music. During the conceptualization phase, this document is surprisingly fluid, with many different people pitching ideas and campaigning for features and story elements. It is not uncommon for this document to be hosted on a wiki or other easily edited digital element to allow for it to be quite agile and open to anyone in the process to make changes and edits. It is during this time that the major parts of the plot, world design, and gameplay are narrowed down. Due to the speed the gaming industry works at, it is likely that the game will actually be in production while many elements are still being designed. A writer may be writing new gameplay segments while artists are producing models and programmers are bringing the game to life.

The final phase of the design document could be called the "fluid design document" phase. Unlike other storytelling fields, the design of a game is never set in stone: a planned feature turns out to just not be fun, technical limitations get in the way of a previously designed element, or (not uncommonly) things are behind and are still being worked out later in the cycle. The design document in this phase will still have everything from the concept doc but some of them will have to be changed due to the aforementioned issues. At this point in the development phase, it is the job of the storyteller to consider all requested changes and figure out how to blend the existing narrative with any new requirements to support the changed elements. This is not always an easy task because revisions at this point in a game are quite costly and some story elements may already be set in stone (anything that required a substantial amount of specialized animation or coding will often be required in the final build, simply due to the amount of resources they took to produce). Here the design document will prove invaluable by giving them access to all the information related to any necessary change, and the ability to quickly add story updates that help make sense of new planned chapters, areas, or enemies.

II. Ways to Get Story into Game

There are several techniques that can be used to graft a story onto a game. Each technique has particular strengths and weaknesses. It is rare to see only a single technique used inside of a game. Many games will use multiple techniques to accomplish their storytelling goals. Some of the primary ways to get a story into game include cut scenes, dialog, exposition, and a few auxiliary methods. As consumers become more accustomed to getting their stories from multiple sources, a gamer may often find parts of the story told via media outside the game itself.

Cut scenes are a common way to get a story into a game. At its most basic, a cut scene is a portion of the game where the gameplay is paused and a small movie takes over to further the story. These cut scenes can range from something as mundane as characters standing and talking to each other to pre-rendered movies that take on a cinematic quality. While other storytelling techniques may also stop gameplay, only a cut scene removes all control from the player, forcing them to watch the scene and wait until the gameplay resumes. Increasingly, cut scenes have been taking on cinematic qualities with advanced camera work, advanced graphics, and pacing. Cut scenes should be considered when a game design calls for an emotional response that can be achieved with the cinematic quality cut scenes are known for.

Another common technique for telling a story is interactive character dialog. In other forms of storytelling, dialog is just character speech. Inside of a game, it will often take on a more interactive form of a dialog tree (Figure 7.9) which occurs when gameplay stops and the player begins to interact with another character or characters. Usually the other character (called a non-player character or NPC) will begin the interaction and wait for the player to respond. The player will be given a set of choices on how to respond to the NPC. Once the player chooses the response they want, the NPC will react accordingly, leading back to another chance for the player to choose a response. This back and forth will usually continue for a few iterations, but dialog can be as short as a one-sided speech from the NPC. In the past, dialog was presented as text but most of the larger games have moved on to using recorded audio dialog. To emphasize the words being spoken and avoid having simply two "talking heads," many games will animate both the player characters and the NPCs during their dialogs. Used correctly, interactive dialog can be a very powerful tool for characterization.

Game characters present the same challenges to a storyteller as do characters in any other format. To make good ones, their design, habits, history, and motivations must all be thought out. Because of the length and scope of many games, it is common to have many more characters in a game than would be present in a movie or book. Because of the abundance of characters, it is important to identify which characters will need to be well-defined (round) characters and which can be less-defined (flat) characters. Like in any other media, rounded characters will need to have histories charted and motivations considered in order to remain believable and interesting to the player for the duration of the game. Round characters in games will include the player character (the one the player controls), other protagonists, and major adversaries. A flat character will typically have a visual design and, if needed, a basic personality. Many flat characters act as background characters that will just be inserted into the game background to make the world feel more "alive." Though many background NPCs have no direct interaction with the player, they are able to interact with the player and they are often given simple responses to player interaction ("It sure is hot out today," a sunbather might say). It is best to treat the flat NPCs more as an exercise in world building, and less in character construction.

For a more dynamic story, it may be necessary to make NPCs who are exceptionally rounded by having them react to player choices. Interaction games like Animal Crossing and the character-driven games made by Bioware contain great examples of reactive characters. Players can interact with these reactive characters in the game through the use of dialog trees and through the in-game actions. As the player progresses through the game, the characters will begin to form opinions of the player based on their dialogue choices and game actions. Should the character's opinions fall too low, they may become cold to the player, or even leave the game. The opposite also holds true: in some, if the player makes the right choices and does the right things, it is possible to develop a deep friendship between the NPC and player character.

Many games are moving towards in-game exposition to deliver their stories. Expositions are all pre-scripted but, unlike cut scenes, in-game exposition does not stop the gameplay, allowing the player to continue the game while advancing the story. The general goal of in-game exposition is to create a world that feels like it is alive by not removing the player from the context of the game like a cut scene would. Occasionally expositions are used to show that events are happening quickly enough that one does not have the time to stop and talk with the other characters. The *Grand Theft Auto* games use exposition well by having NPCs speak to you about mission objectives while the player is driving them to a destination. Many games use exposition to help with characterization by triggering background conversations during gameplay. These background conversations can be between any of the characters in the game and are able to provide small bits of characterization while the player may be doing an otherwise fairly mundane action.

Many games will provide non-essential parts of a story through easily missed or overlooked elements, relying on a player to do some extra work or exploration to find them. Because they are hidden and tough to find, it is difficult to tell a game story with only these elements (though there are a few that do it well). Games that do have findable items or database entries usually use these hidden elements to explain some of the background of the game world and its history.

If a game's story is written and presented well, many players will want more. It's important to keep in mind that a game's story does not all have to be revealed just inside the game. As people get used to taking in content in a multitude of ways, it is not uncommon to reveal parts of the story on a game website, inside the game packaging, or on a social media site. These are great places to develop back-story or flesh out characters that simply could not get screen time within the game. A good example of this is companies who write about updates to their "live" games (online games like Star Wars: The Old Republic and multiplayer shooters) from the perspective of a person or company within the game world itself, trying to keep players "in the game."

III. Interactive Structures

Once a storyteller has decided on a story and how much agency they want to give to the player, they will need to decide on an interactive structure to best control that agency. Some interactive structures naturally provide a user with a great amount of agency, while other structures can be quite limited It's important to keep in mind that many games do not use just one structure; instead they will combine structures to form a larger interactive system, using each structure to its strength during the course of the game.

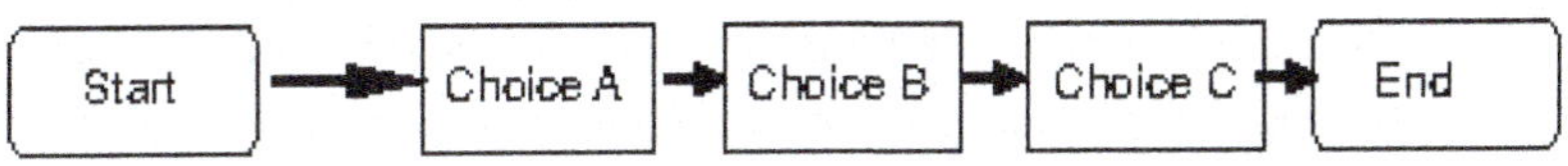

Fig 8.5

Linear game design flowchart

The most basic form of interactive structure is the linear structure (Figure 8.5) because it has a fixed beginning, middle and end. This structure most closely resembles a traditional story. In order to engage the player, a game employing the linear structure provides brief moments of interactivity during which the player is allowed to take a number of small actions until they find the choices that move the gameplay forward. The game will progress through many of these sections, each leading towards the end of the game. This is a preferred structure for storytellers with a story to tell that does not allow for any deviations. The "Phoenix Wright" series from Capcom is a great example of this linear structure. During court cases, there is only one correct answer in any interactive segment to progress the case. An incorrect answer will simply offer the player a chance to choose a different argument.

Another option for the storyteller is to give the players an open space to explore filled with many different interesting things to do. This is commonly called an "open world" or "sandbox" game where, much like real life, the player is not given any direction on what to do or where to go. Typically this sort of game will have some history provided to frame the nature of the world and will tell a story through the characters and items found within the world, with very little story actually being forced on the player. Instead of providing a story, it is expected that a player in an open-world game will create a story for themselves through their actions and the things they encounter in the game. The game Skyrim, after its opening moments, drops the player into a world with very little direction. Someone playing Skyrim might explore a few caves, help some people with their shop, and make some swords during a single gameplay session. As they play the game, they will come to understand the world through the places they visit, the people they interact with, and books and documents they pick up and read along the way.

A funnel structure (Figure 8.6) is a common way of structuring a story that is seen very often in dynamic dialog trees. Funnels are a linear structure with choices that can lead to different results. In this branching structure, every choice can lead to a different version of the story that must be considered and scripted out. For example, in a game using the

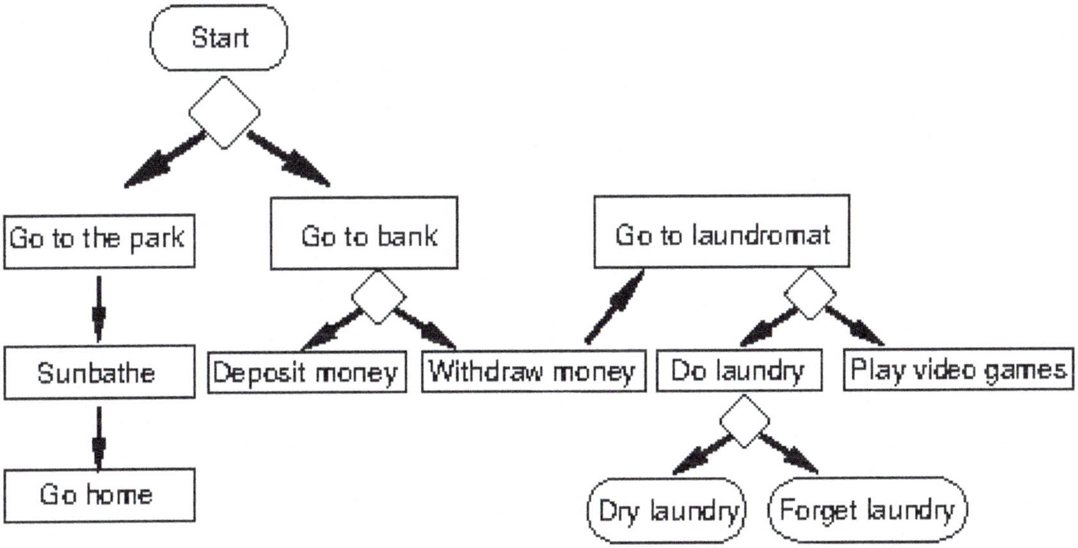

Fig 8.6

Funnel flowchart

funnel structure, there might come a point when the player is asked if they want to go to the bank or Laundromat. In this game saying "yes" would lead to a game state where the player heads to one of these places, and saying "no" would lead to a world where the player character sits on the couch at home, wishing he/she had gone to the bank or Laundromat. If more choices are presented once the game begins, the story can quickly lead to very different places. Funnels can be fun for the player by tricking them into feeling their actions can really change the game world. The downside to a funnel type games is that writing for all the different paths can be challenging, requiring the storyteller to make multiple stories and keep in mind all the decisions a player has made leading up to any particular story segment. Sometimes it is impossible for a storyteller to write all the different paths that a particular funnel may take. In order to mitigate these issues, some options will lead back to previous choices, or lead to the same game state, regardless of the player's choice.

Sometimes a storyteller may want to give the player the illusion of freedom, while still driving them towards a predetermined conclusion. When using the pyramid structure Figure 8.7), the storyteller gives the player an array of tasks to complete ("wash the car," "take out the trash,"

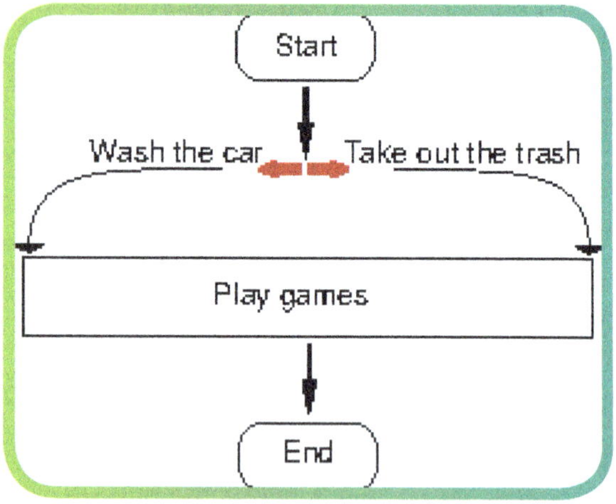

Fig 8.7

Pyramid flowchart

and "play video games," for example). The player may complete these tasks in any order they choose. Perhaps the player may choose to "take out the trash" first, leaving them to still finish the other items on their list. If they choose to "wash the car" next, they will be left with just "play games" as their final task. Once a player completes all the tasks assigned to them, they will be presented with an unavoidable story point (perhaps after all their work they inevitably fall asleep). By using the pyramid structure, a storyteller can give the player the illusion of choice, while still leading them into a predetermined finale.

A storyteller may wish to have part of the freedom of an open world while also being able to have the restriction of a linear structure. The hub structure (Figure 8.8) provides a solution by giving the player a central location that enables them to choose what part of the story they wish to experience next. The hub can come in many different forms: a list of options, a series of doorways, or a collection of icons to click on. Each of these portals will lead to some interactive segment that will likely make use of one of the other interactive structures. Once the player completes their chosen story section, they are then returned to the hub and prompted to choose another portal. This type of structure has been used successfully in games like Super Mario 64, the Mass Effect trilogy, and many others.

It is not uncommon to see only a single interactive structure used within a game design as illustrated in (Figure 8.9), a combination of hub and funnel. Instead, most games will use some combination of

Fig 8.8

Hub flowchart

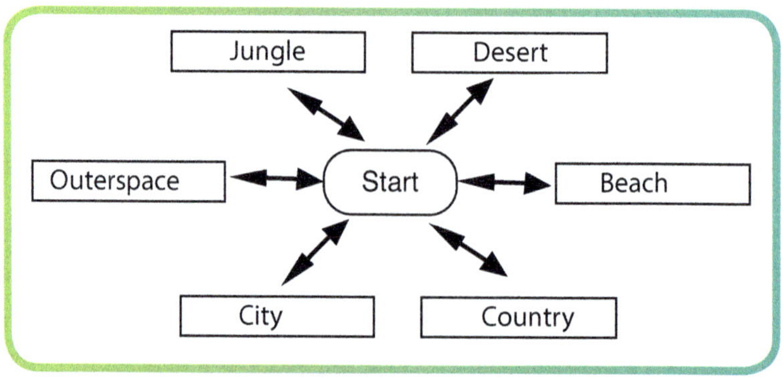

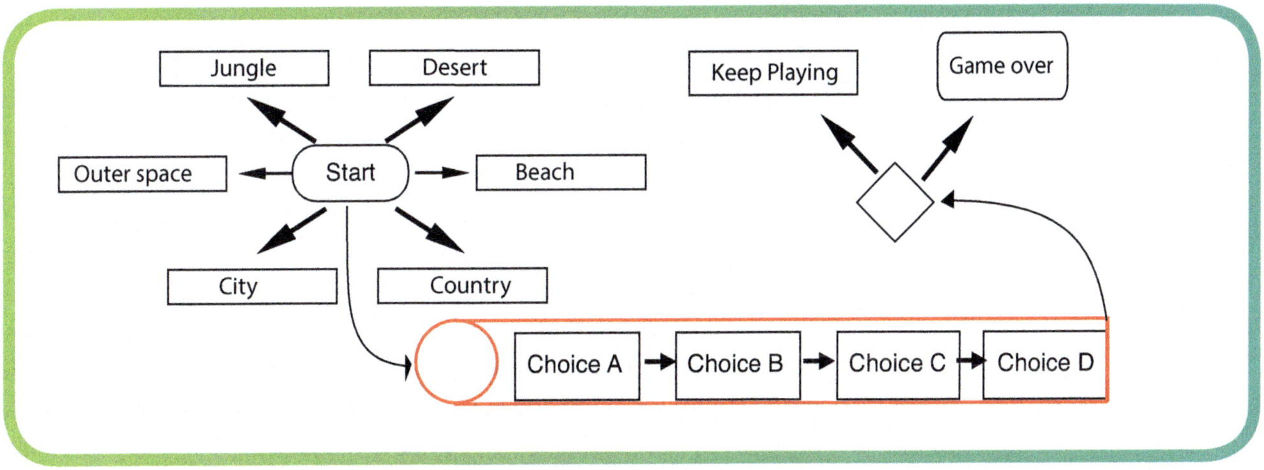

Fig 8.9

Combined flowchart

structures in order to achieve the right balance of agency and story pacing. One particular story may call for the player to have latitude in order to do a lot of adventuring until they complete a set number of tasks in an open world. Once they finish all those tasks, the player will then be taken to a plot segment with a predetermined outcome that is reached by leading the player through a series of right or wrong choices. This example is actually a combination of three different structures: open world, pyramid, and linear. Each structure has its own strengths and weaknesses, and care should be taken to choose the right structure for the desired effect on the player. For example, if there is only one right answer in a story, it would not make sense to give the player the freedom of a funnel or open world. Alternatively, if the goal of the game is to immerse the player in a vivid world, a linear story does not make sense either.

Unfortunately, textual descriptions of interactive structures cannot communicate very well the particular "feel" each imparts. For those who are reading about these structures and have never played through a game before that makes use of them, it would be helpful to pick one or two modern big-budget (or, "AAA") games and play through them. After a few hours in these games, the strengths, weaknesses, and use-cases of the different interactive structures will become apparent and one will be able to tell when a game is intentionally restricting a player's choices and when it is giving a player quite a lot of options and freedom.

IV. Finding a Game Genre

There are many different genres of digital games that are played in very different ways. Some of these game genres are more capable of supporting stories than others.

The following will be a brief (and non-exhaustive) discussion of some game genres. As the gaming industry matures, these genres have started to overlap, making it difficult to classify some games as "just a puzzle game" or "just an adventure game." A game that spans genres may offer storytelling tools that a single genre does not. In order to tell the best story, it is important to always research exactly what elements are going into the game and story before accepting that the game falls into its overarching genre.

Puzzle games are a fairly simple game genre that typically involves the player doing a bit of thinking before taking a single action. Some great examples of puzzle games include Tetris, Bejeweled, and, for a non-digital example, Where's Waldo. It is rare to see a story associated with a puzzle game. When a story is included, it is typically used to introduce new puzzles to the players through the use of NPCs who either ask for your assistance or require you to complete their puzzle before letting you pass. Mystery stories appear to lend themselves well to the puzzle genre due to the already overlapping elements of mystery-hypothesis-solution found in both the story and the game.

The strategy genre also requires more thinking than actual action. Strategy games typically consist of a set of resources to use, a goal to overcome, and opponents vying for the same goal (often through control of the resources). Common strategy games involve moving military units around to defeat an opposing army, though that does not have to be the case. Other games that could be called strategy games include chess, go, and Settlers of Catan. These games are surprisingly good at telling either actual or alternate histories by giving the pieces in the game specific roles and names (the board game Risk is a great example). When telling a story using a strategy game, it is common to progress the story in between each play session through the use of cutscenes and dialog. These cutscenes are used to set up why the player is going into the next play session and whom it is they will be dealing with when they enter the next match.

Games that can fall under this genre include Mario, Zelda, and Halo series. These games typically focus on what some would call "twitchy" gameplay, which is gameplay that requires fast reaction times to address opponents and evade dangers. Although many assume these kinds of games cannot be good mediums for a story, there have been several that have used the genre to tell memorable stories. The plots of Bioshock, Mass Effect, and Halo have all received good reviews, and many gamers look back with fondness on the stories of many Mario and Zelda games. The stories in action games tend to be very linear, though there are some exceptions (such as Mass Effect). The previous discussion about balancing and reinforcing the link between the challenge of the game and the story the game tells applies very strongly to action games. If the story and gameplay are balanced well, the game player will remember the game not just for how fun it is, but also for the story they experienced along the way.

A final genre that presents a lot of opportunity and challenges for a storyteller is the role playing game (RPG). Games in this genre give the player a setting and a character whose role they are expected to play faithfully. Some RPGs will force a particular role on the player (they are always the space marine, though many others will let the player choose a role they find interesting (perhaps between a baker, firefighter, and grocer). During gameplay, players take actions and deal with NPCs in a way that is consistent with the roles they have been given. Choosing certain options affects how the rest of the game plays out, giving the player a sense of really mattering in the game.

The Dragon Age series does an excellent job using player choices and dialog: two players can have very different play through of any of the games based on the choices they make. A player might not find all the same NPCs as another player, their character may or may not be alive by the end of a game, or many other small and large things may be different in their games.

Until a good storytelling artificial intelligence can be built, all this content must be designed and written by an author. This means the

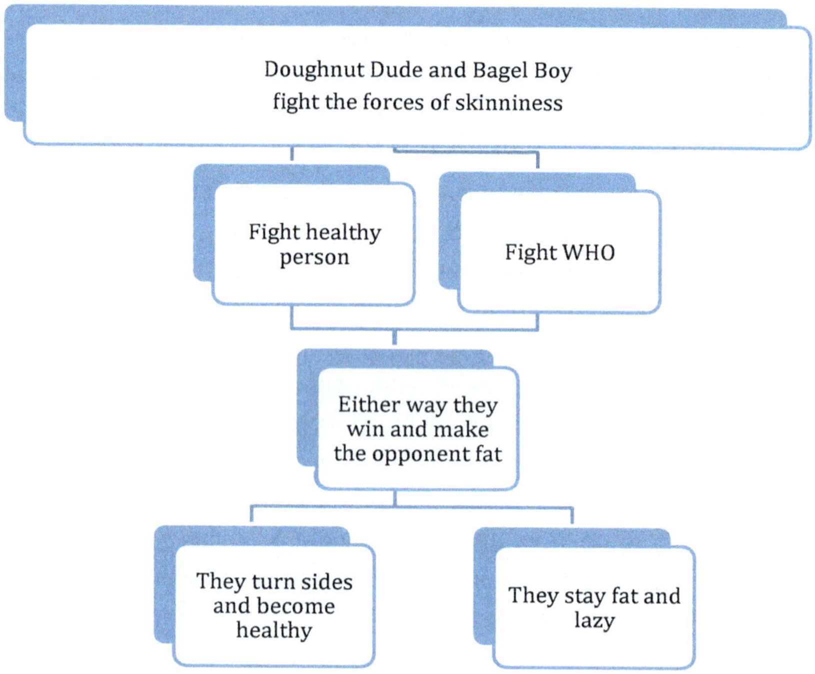

Fig 8.10

Choose Your Own Adventure (CYOA)

storyteller will need to work extremely closely with the designer in order to keep all sorts of variables in mind when writing, along with making sure that every plot variation is considered and fully defined. It is not an easy task, but can create some very compelling (and "replayable") gameplay when done well.

C. LOOK AT OTHER INTERACTIVE STORIES

Other types of interactive stories are *The Johnny Cash Project*. This is a global art project based on the song by Johnny Cash, "Ain't No Grave." This type of interactive story invites end users to download a still image from the music video and create their unique interpretation of that picture. The image is uploaded to the site alongside other participants. The images are strung together into a series that creates a living homage to the late Johnny Cash as walks the railroad tracks (1932–2003). *The Way Beyond Trail* is an interactive story created as a commercial to promote the 2007 Jeep Patriot. The story centers on a trip across America with forty-four story modules that weave an entertaining digital story in which the audience steers the plot. The interactive installation work of Camille Utterback and Scott Snibbe combine technology with art. These artists display their work in public places that invite participation.

D. CHOOSE YOUR OWN ADVENTURE

Creating a video that is interactive is a great beginning and easily accomplished. In this is example of a *Choose-your-own-adventure* (CYOA) story you should also consider having at least 4 points of interaction in which an end user can choose what to do next. You can *create the annotations* in YouTube (Figure 8.10).

E. ROLE-PLAYING AND MMORPG GAMES

Role-playing and ARG stories (Alternative Reality Games) attempt to craft a story that is believably taking place *in our current world*, typically using multiple mediums—audio, video, interactivity—and are almost always puzzle based. Most ARGs are time limited (run for a predetermined amount of time) and they blur the lines between fact and fiction. Techniques used in ARGs are players become actors and connect through social media, gameplay, and mobile technology. Typically ARG stories are staged such as zombies overpowering the world or a treasure that is stolen with clues to find it. ARGs use GPS, email, text messaging, and other forms of communication to find the physical object. Sometimes the games are staged events (fight at game conventions, "theft" of a car). Most start with the rabbit hole that draws in the players. There is no town crier or someone directing the game, but a release of little clues that, with a bit of searching, uncover more puzzles to be solved. The *ghost story* is an example of an ARG that calls and sends text messages to players. ARGs are not all entertainment; they can be used for educational purposes as well.

Massively multiplayer online role-playing games (MMORPG) combine role-playing in a large group of players who compete in a virtual world. The main distinction between role-playing games and MMORPG is the number of players and the global accessibility. Some games are free while others cost to play.

I. Physical Interactive Games

Interactive stories that involving physical exercise-type games are treasure hunts such as geo caching. This is a global interactive treasure hunt that involves finding the object using GPS and leaving it or replacing it with another object. The player must log into the website in order to play or create their own geocache. The Fun Theory is a design firm that uses imaginative methods to change people's behavior. By creating piano stairs in a subway station in Stockholm, Sweden, the firm encourages people to walk the stairs rather than taking the escalator. Interactive stories can bring people together for a variety of purposes. Games such as Zombies vs. Humans and Flash Dancing are examples. This can be a face-to-face experience or a globally orchestrated event.

EXERCISES FOR STUDENTS

a) Discuss agency and pacing in a story and find examples.
b) Discuss types of navigation.
c) Discuss interactivity and find examples.
d) Discuss games and types of interactive structures.
e) Discuss role-playing, ARG and MMORPG games and find examples.
f) Work in a group and develop an interactive story with 7 points of interaction. (Choose your own adventure CYOA interactive digital story.)

Part Four

4 WORKING IN THE INDUSTRY

9. OVERVIEW

A. WORKING IN THE FIELD AND FREELANCING

Now that you have the digital story completed, what do you do with it? There are a multitude of platforms to publish your work online including social media. I have heard the comment that it's difficult to get hired without experience. If you want to work independently, start by doing research about self-publishing opportunities. If your story is time-based, there are video social media sites such as YouTube and Vimeo. Social media is a way to have your story read or viewed and also a way of getting feedback. Consider transmedia storytelling that takes advantage of multiple platforms and formats and gives your story a wider appeal, and, if your story becomes viral, longevity. Try entering any local, regional, national or international storytelling competition. National Public Radio's (NPR), three-minute story competition is an example on a national level. Look into community-based writers' workshops, residency programs, and writing articles for your community paper. If you are writing a script that you plan on producing, seek actors at the local college or university and funding from sites like Kickstarter.com. The question is what part of digital storytelling interests you and what do you see yourself doing? Stories are created for many purposes. To break into the business of digital storytelling, you need to be realistic about the competition and what the culture is like. Research the industry, not-for-profit and commercial, and if possible visit. It's all about connections.

Freelancing is a way to make a living as your own boss by taking short-term and long-term jobs. There are many parts of freelancing that you should become familiar with. Consider the parameters of the freelance job and learn how to be flexible because each job might be very different. Think about how the client is involved and work out the terms of the contract. What are the time restraints, the deadline, and the salary? If you have no experience, you might take a business course. Each company has a unique culture that you as a freelancer need to fit into. Leon Nowlen [9] special effects artist and former student, has been freelancing for many years. His portfolio includes working feature films, motion graphics and animations. Leon shared some advice with me to pass on to students. It is most important to learn the concepts and skills of good digital storytelling because the software and technology will continually change. Leon also made the point that freelancers need to build a network and maintain good rapport with co-workers and clients. You never know where the next job will come from or who will have recommended you for it.

B. CREATING A PORTFOLIO

In order to promote your digital story, you need to have a website. On your website should be your best work, resume, skills, education, honors and awards, experience, and contact information. Your website needs to be attractive, easily navigable, and make the end user take notice. It is important to mention your individual contribution if you are including a team project. Check out other websites with similar purpose. Make sure you create and update an online business social media profile such as LinkedIn.

C. CREATING A DEMO REEL

If your interest in digital storytelling is with time-based media (animation, film, video, audio, motion, and stereo graphics), then you will need to create a demo reel. The reel needs to show your best work and what you did if it was a team effort. Keep the demo reel short. Pixar requires it to be under four minutes. Rule of thumb as in a portfolio is to include your best work. Never use copyrighted material unless you get permission. As mentioned in Audio/Video/Production: pay attention to technical details such as the resolution and aspect ratio; size and quality. The demo reel needs to have your name and contact information at the start and end of the footage. Include it in your website and social media. Include any visuals such as storyboards, illustrations, animatics, and script if applicable. If you are required to send your work as a DVD, make the cover for the DVD and case attractive and have all your contact information printed on both. Include a cover letter, resume, and website.

D. VIDEO GAME ADVICE

Probably the most important thing you will write will be the cover letter. This document will most likely be the first and possibly the only one read before the demo is played. It has to tell the publisher everything about you and your proposed game. A cover letter is typically a one-page document with an introduction, a body, and a conclusion, usually about four solid paragraphs that sum up the entire game proposal.

The main points you want to get across in the cover letter are that you have a great game idea (marketable) and that you are able to make that idea a reality. The letter very briefly sums up all that is contained in the proposal and mentions specifics only if you have a hot selling point such as a cutting-edge technology, license, or top-name talent on board. And this is not just a document that sells the game, but you as well. The reader of this letter will notice bad formatting, spelling errors, and how well your thoughts are organized, among other details that will speak of the letter's author.

When writing the cover letter, remember to consider your audience and state why you want to be published by this particular publisher. If possible include a reason you are a fit for this publisher. If, for instance, you have read an article where the publisher has been quoted as saying that they are looking for your type of game, mention it here. And remember to conclude your letter with a request for action. Don't just say thanks and goodbye; ask the reader to follow up or better yet, tell them you will be following up.

I. The Game Overview

The game overview should contain the basic data about your game. This generally includes the following:

Game Title. Be sure to indicate that your game's name is a "working title," as this means that you are aware that the title may change. Clinging to the name you gave your game is a sign of inexperience, and publishers have may have legitimate reasons for changing the name such as copyright issues, marketing reasons, and (believe it or not) the possibility that the title you came up with is simply not the best. This doesn't mean you shouldn't give any thought to the title, though. You should have a title that speaks to the publishers and connotes the essence of your game.

II. Category or Genre

Usually publishers want innovation, not invention. That means that the preference is for an improved version of the latest hit, not a new, untried idea that nobody can define in traditional market terms. Games based on proven genres (action games or RPG games, for example) are easier to market and easier for retailers and gamers to understand.

In the book *Game Architecture and Design* by Andrew Rollings and Dave Morris, genres are defined in their truest sense and can be broken down into about seven types:

- Action: Lots of frantic button pushing.
- Adventure: The story matters.
- Strategy: Nontrivial choices.
- Simulation: Optimization exercises.
- Puzzle: Hard analytical thinking.
- Toys: Software you just have fun with.
- Educational: Learning by doing.

Of course genres can be defined, like Action/Adventure titles and Action/Strategy and so on. On top of these classifications you pile on the details about your technology, marketing, and other factors, and soon these traditional genre labels become less effective. Conveying details about your game's technology, gameplay, and other critical information in just a few sentences can be extremely challenging.

For example, much confusion surrounds the simple term "3D." If a game describes itself as "3D Action" on its box, what does that tell you? Will it be first- or third-person perspective? Will there be multiplayer support? Obviously the term "3D Action" doesn't answer all the questions someone could have about a game. So you're back to using a list of genres, around which you'll need to provide some context. In other words, don't just put "3D Shooter." If the publisher has a particular negative bias towards that term, chances are your game won't get the second look it may deserve.

When discussing your game, keep the genre in mind, as it is a foundation for communicating your vision to others. Once you have a clear idea of your game, "It is a first person shooter with shades of military simulation and strategy in the multiplayer mode," you can proceed to describe it in visual terms on paper, eventually breaking it down into the elements that will comprise the design document.

III. Technical Specifications

The technical specification explains the how and what of the technologies and tools your team will use to develop the game. This section forces you to consider all of the tasks that are needed to complete the game development process. The technical specification is extremely important for the proposal as well as the development schedule, but it's beyond the scope of this article to go into depth about it. For more information see "The Anatomy of a Design Document, Part 2: Documentation Guidelines for the Functional and Technical Specifications" by Tim Ryan (www.gamasutra.com/features/19991217/ryan_01.htm).

IV. Minimum System Requirements

List the platform, hardware, software, special peripherals, RAM, and so on that will be needed to run the game. You may want to demonstrate your understanding of the gaming market in this section by citing the current installed base of this certain PC configuration.

V. Media

If you need four CDs or a DVD, then discuss it in this section—and justify the added expense the publisher will see in terms of replication, packaging, and even design fees. Explain why that media is justified and why it will make the publisher money. Keep in mind that the publisher generally wants a game that will be economical to make, easy to market, and fits current standards of shelf space and box design.

VI. Internet/Multiplayer Capabilities

Clearly define your game's single and multiplayer modes. If you are lacking one of these modes, you must present a solid argument why the market will support a one-mode-only game.

VII. Development Stage

Each publisher has its own definition of the development stages, and you should understand these stages when you work with a game development company. You need to indicate to the publisher what stage your game is at. You may want to stay away from labels and simply describe how far along your game is and how much farther it needs to go. If you indicate that you are at beta stage out of ignorance or misunderstanding when the publisher judges the demo to be at alpha, you could inadvertently give the publisher the impression that you have a half-functioning game when you consider it almost finished.

Here are the generally accepted definitions of game development stages:

- Work in Progress: This is pretty self-explanatory. You have something done, maybe some art and a little code. Nothing is expected at this stage but a firm understanding of where you are headed and the demonstration that you can get there.
- Alpha: The game is running to some degree at this stage. User interface is defined, the general layout of the program is set, the look

and feel have been achieved, and the programmer is just starting to get the cold sweats as he realizes how much he has left to do. At the alpha stage you should be able to demonstrate the gameplay and the look and feel.

- Interim Beta or Second Alpha: This comes before the final beta stages. Some bugs and errors have been found and fixed. Your game is essentially running and mostly done. Some tweaking is taking place and initial beta testing is starting. Artists are having gag reflexes at the sight of the game and will often break down in tears. This is the stage publishers most want your title to be at when they look at it in a proposal.
- Final Beta: All features are functional and complete. Things like the installation,, help files, and splash screens are complete. All text files have been checked for spelling and grammatical errors. The development team has tested the game and all of the bugs found during that testing have been corrected.
- Gold Master: This is the final version of the game. It has been mastered for duplication and dissemination.

E. THE GAME TREATMENT

> "I didn't have time to write a short letter, so I wrote a long one instead."
>
> — Mark Twain (1835–1910)

This quote sums up a very important aspect of the game treatment: it is often easier to write a ten-page document describing your game than to write one page that says everything concisely and effectively.

The game treatment is your primary selling tool that quickly orients the publisher to your game (genre, platform, story, and other elements should be mentioned). The treatment is your opportunity to convince the publisher that your idea is as complete and crystal clear as you believe it to be. If you can't briefly explain your game idea, then the following might be true of your game:

- It is too big and complex. Your game idea may be too ambitious.
- The game is not as good as you think.
- You need help crystallizing the core aspects of the game. If you think this is the case, get help from an objective source who knows something about games.

One approach to writing the game treatment is to cite your game's "unique selling points" (USPs). USPs are aspects that differentiate your game from the competition, offer gameplay value, and ultimately make consumers want to buy the product. You should be able to determine the USPs of your game after writing the competitive analysis section of your proposal. If yours is truly a good idea, the publisher may even read your design document all the way through.

The game treatment should also grab the publisher's attention by focusing on the bottom line. You can do this by citing sales projections based on solid market and sales information relevant to your game.

When you have distilled your game into a single game-treatment page, have others proofread it for you to see if they understand what you are trying to say and find the game idea compelling. Have it edited and reviewed by as many qualified people as possible.

Note: Many people erroneously believe the treatment is the first thing written, since it is usually the first thing read. But this document should be written last, right before you approach the publisher. The reason for this is that at this point you know as much as possible about your game. The treatment is a distillation of all the work that you have done designing, researching, and otherwise developing the game on paper.

Publishers and agents, to weed out developers and game ideas, often use proposals and forms. Not just bad developers and game ideas are weeded out, however; publishers are just as likely to weed out inappropriate or unwanted ideas and developers.

F. THE COMPETITIVE ANALYSIS

The competitive analysis illustrates to the publisher how you stack up against your competition. It must explain why your title will outsell the competition, yet how it will be similar enough to be sold right next to the competition on store shelves. A simple way to graphically illustrate how your game stacks up against the competition is a table that lists functions and features down the X-axis, with your game and its competitors across the Y-axis, checking off the features that each game has. You can also list other competitive differentiators here, too, such as licenses, technology, and development costs—anything that will make your title perform better in the retail channel.

Be prepared to discuss each point of your analysis as it relates to the competition. For example, if you list a great technology developed in-house as a selling point, you need to be prepared to discuss why it will give you an edge over the competition. Will it save development time or money? Will it make the game noticeably better? Or is it just a source of personal pride that doesn't translate into publisher benefit?

The competitive analysis should contain at least five products (in today's marketplace, you shouldn't have a problem listing ten or more) and at least five features of those products with a few paragraphs discussing the competitive analysis in terms of how you drew your conclusions. The most important determinant of what your competitive analysis contains is your game idea. You need to include your real competition and not leave out a top-selling game, hoping the publisher won't notice your omission. The same goes for features: you can't leave a feature out of your analysis because your competition has it and you don't. Most importantly, make sure the features you include have a selling point to them. And be honest, too—there will be games that simply have features going for them that yours won't have.

The competitive analysis requires you to market your game—and yourself. You are being given the chance to tell the business guys that your game will perform in the marketplace better than titles X, Y, and Z for reasons that you state as persuasively as possible in their language.

Done correctly, the competitive analysis can help you jump ahead of other developers by demonstrating to the publisher that you understand its goals. And once you've written it, you'll be able to rattle it off in conversations—and that helps convince others that your title stands a strong chance of performing well in the marketplace.

Conclusion

The craft of digital storytelling is to use the theory, concept, principles and development process provided in this book in such a way that the audience sees and feels what the author wants to convey. The goal of digital storytellers is to draw the audience into the story unfolding the narrative to keep the audiences attention. From the beginning of history to the present, humans have been curious, creative and storytelling has been a universal way to express their ideas. The purpose of digital storytelling whether it be formatted as an animation, game, interactive encounter, film, video, ebook, audio recording or performance is too educate, entertain, inform and transcend the story to create a compelling experience for the audience. The challenge of a great digital storyteller is to begin with an interesting concept and using the information found in this book, bring your story to fruition. Storytelling is a combination of research, imagination and good writing skills. Stories are developed as Freytag's three act structure that includes POV, character development and digital media to enhance the narrative. Becoming a good digital storyteller takes practice and patience. Through the course of reading this book I hope you have gleamed a greater understanding of developing a digital story for any media. As you move forward in your digital storytelling development continue to read, listen, view, and experience life. Become an observer/listener and keep a digital journal of interesting events and characters. At the end of the day make the story interesting, contextually and multimedia rich, visually compelling and worth someone's time to view, read or engage with. Write a story that makes your audience care. Good Luck.

EXERCISES FOR STUDENTS

a) Research places of employment that interest you.
b) Create a portfolio and website.
c) Write a cover letter and resume.
d) Conduct a class critique of portfolio, cover letter, resume and website.
e) Develop a promotion plan.
f) Write your digital story and submit it to competition.

REFERENCES

[1] http://en.wikipedia.org/wiki/_storytelling
[2] http://en.wikipedia.org/wiki/Literary_adaptation
[3] http://en.wikipedia.org/wiki/Charachetypecter_arc
[4] http://en.wikipedia.org/wiki/Digital_convergence
[5] http://en.wikipedia.org/wiki/
[6] http://en.wikipedia.org/wiki/Steven_Sasson
[7] http://en.wikipedia.org/wiki/Personal_digital_assistant
[8] http://henryjenkins.org/2011/08/defining__further_re.html
[9] http://www.imdb.com/name/nm2996811/

CPSIA information can be obtained
at www.ICGtesting.com
Printed in the USA
LVOW02s2338230816
501571LV00016B/322/P

9 781626 614109